EUGENIC STERILIZATION

EUGENIC STERILIZATION

Compiled and Edited by

JONAS ROBITSCHER, J.D., M.D.
Henry R. Luce Professor of Law and the Behavioral Sciences
Emory University Schools of Law and Medicine
Atlanta, Georgia

CHARLES C THOMAS • PUBLISHER
Springfield • Illinois • U.S.A.

Published and Distributed Throughout the World by

CHARLES C THOMAS • PUBLISHER
Bannerstone House
301-327 East Lawrence Avenue, Springfield, Illinois, U.S.A.

ISBN 0-398-02699-8

Library of Congress Catalog Card Number: 72-88489

With THOMAS BOOKS careful attention is given to all details of manufacturing and design. It is the Publisher's desire to present books that are satisfactory as to their physical qualities and artistic possibilities and appropriate for their particular use. THOMAS BOOKS will be true to those laws of quality that assure a good name and good will.

Printed in the United States of America
R-1

CONTRIBUTORS

JONAS ROBITSCHER, J.D., M.D.

Henry R. Luce Professor of Law and the Behavioral Sciences
Emory University Schools of Law and Medicine
Atlanta, Georgia

GEORGE TARJAN, M.D.

Professor of Psychiatry, Schools of Medicine and Public Health
Program Director, Mental Retardation, Neuropsychiatric Institute
University of California at Los Angeles
Los Angeles, California

JULIUS PAUL, Ph.D.

Professor of Political Science, Department of Political Science
State University of New York
Fredonia, New York

FRANK J. AYD, Jr., M.D., F.A.R.A.

Psychiatrist
Editor, The Medical-Moral Newsletter
Baltimore, Maryland

ALAN F. GUTTMACHER, M.D.

President, Planned Parenthood World Population
New York, New York

DONALD GIANNELLA

Professor of Law
Villanova University School of Law
Villanova, Pennsylvania

ALYCE McL. C. GULLATTEE, M.D.

Assistant Professor of Psychiatry
Howard University College of Medicine
Washington, D.C.

MEDORA BASS

Counselor, American Institute of Family Relations
Consultant, Association for Voluntary Sterilization
Santa Barbara, California

PAUL S. MOORHEAD, Ph.D.

Associate Professor, Department of Medical Genetics
University of Pennsylvania School of Medicine
Philadelphia, Pennsylvania

PREFACE

THE idea for this book on eugenic sterilization was born in 1970 when a physician in Pennsylvania sought advice on whether he could sterilize a mentally defective young girl who might become promiscuous. Research indicated that in the absence of a statute authorizing the procedure, the sterilization would probably be illegal in the state, but it also indicated the answer might be different in many other jurisdictions. Of equal interest, the research revealed a scattered, somewhat outdated literature on the subject – there was no one current comprehensive work focusing solely on eugenic sterilization

In order to revive interest in this topic – for itself and also as a prototype of other topics involving biomedical interventions on unwilling patients – a meeting was convened to be held in connection with the annual medical and psychiatric meetings of the state societies. One main purpose of the program on eugenic sterilization was to augment the literature so that state legislatures, which frequently have introduced bills authorizing eugenic sterilization or bills repealing existing statutes, which authorize such sterilization, could have the benefit of the meeting's transcript as a kind of *legislative hearing* on eugenic sterilization.

The meeting was held on November 18, 1971, at Lancaster, Pennsylvania. This volume is not a transcript of that meeting; instead, it is a collection of papers by the participants who were asked after the meeting to revise and formalize their contributions. I hope that the resulting work escapes the flaws of the printed symposium which may sound fine but does not always read well.

The choice of participants for the meeting was a problem. Originally the co-chairman and I thought that every point of view could be represented, liberal and conservative (whatever those labels mean in this context), psychiatric, legal, the field of retardation, Catholic and non-Catholic, ethnically-major and ethnically-minor. We soon concluded that if every point of view was represented by a protagonist and an antagonist, the program would go on for more than its one day. We therefore arranged for a conference of nine authorities with strong positions and/or wide knowledge in the field. Julius Paul was invited because he had studied the political and sociological implications of compulsory sterilization. Medora Bass was invited because she had crusaded for more widespread use of eugenic sterilization, and Alan Guttmacher because of his concern over decades with the problem of population and reproduction and his leadership in the field of population control. We had planned to invite a spokesman for the Catholic position; when we consulted Frank Ayd for help in finding the right representative he offered to double as a psychiatrist and as a spokesman for the Catholic position. The panel needed a black representative; it had only one woman member and could use another, and we were fortunate enough to secure the cooperation of Alyce Gullattee who could speak as a woman, as a black, and also as a psychiatrist. To put the problem in legal perspective, Donald Giannella, Professor of Law at Villanova University School of Law, was invited; he had previously written thoughtfully about abortion. To keynote the program, George Tarjan, psychiatrist, whose work with the retarded filled many pages of a distinguished curriculum vita, seemed an ideal choice. After the panel had been announced and as the conference time grew near, we belatedly recognized that the program could not go on without a geneticist; Paul Moorhead agreed on the stipulation that, because of the lateness of his invitation, he would not speak as formally (or as lengthily) as the other participants.

While the speakers could not represent *all* points of view, they could represent a wide range of views and a broad experience – Catholic, non-Catholic, black, white, retardation specialists, psychiatrists, lawyers, males and females, pro and anti, scientist and humanist.

I want to thank all the participants for their original contributions and for their rewriting efforts. Special thanks go to Julius Paul for preparing appendices to this volume, including a listing of up to date legislation for all 50 states.

My appreciation is also extended to the three sponsors of the meeting, the Pennsylvania Psychiatric Society, the Pennsylvania Medical Society, and the American Medical Association and to two companies who defrayed the expenses of the meeting, Sandoz and Merck, Sharp and Dohme. Among the many helpful people involved were Paul J. Poinsard, M.D., President of the Pennsylvania Psychiatric Society, my co-chairman of the Program Committee, Ulysses E. Watson, M.D., and the committee's members, Robert Sadoff, M.D., and Norman Jablon, M.D., and the able Executive Secretary of the Pennsylvania Psychiatric Society, Terry R. Lenker.

The preparation of the manuscript was supported by PHS Training Grant No. 2-T01-MH-12,555-02, NIMH (Center for Studies of Crime and Delinquency).

I wish to acknowledge the editorial direction given by Dr. Ernst Jokl and my heartfelt thanks go to my wife, Jean, for her consistent support and to my Administrative Assistant, Naomi Dank, who saw this work through publication and provided sound suggestions and hard editorial labor, and to Barbara Brown who so ably handled the secretarial duties on this project.

JONAS ROBITSCHER

CONTENTS

EUGENIC STERILIZATION

Chapter I

EUGENIC STERILIZATION: A BIOMEDICAL INTERVENTION

JONAS ROBITSCHER

EUGENIC sterilization is a biomedical intervention that has given us eighty years of experience with the legal, moral, and philosophical problems that arise when society decides to use medical means to impinge on the rights of some people in order for others to benefit.

Today we have a host of such interventions used or advocated – enforced therapies such as electroshock and chemotherapy used for mental patients; biogenetic engineering approaches, including prebirth predictive techniques to determine candidates for abortion; organ transplantation from living donors; the rating of patient priorities for dialysis; and – most recently – cloning. Eugenic sterilization arguments have been the prototype of arguments made for and against newer interventions and determinations.

Eugenic sterilization – defined as the improvement of the quality of the population by means of controls on child bearing based on eugenic principles – does not awaken feelings of immediate threat, or, for its proponents, of immediate promise, in most minds. We do not know many people who in the present are being, or in the near distant future, will be eugenically sterilized. Only an occasional news item in which the forcible sterilization of the poor or the unfit or the immoral is advocated serves to tell us that eugenic sterilization is still a lively topic (although not as lively as a related topic, the control of population that has come

NOTE: This study was supported by PHS Training Grant No. 2-TO1-MH-12,555-02, NIMH (Center for Studies of Crime and Delinquency).

to the center of the stage as eugenic sterilization is being discussed less frequently).

Old and unfashionable, the topic of eugenic sterilization still is worth discussing; perhaps it is especially worth discussing because it has been around for so long, because it is decreasingly utilized and so the heat is somewhat out of the issue, and because it is a more easily encompassed issue – smaller and more easily explored – than recent related topics. The losses and gains of the intervention are easily counterposed, on the one hand, the right of those to reproduce whom society sees as eugenically unfit, on the other, the right of society to decrease its burden of children born inferior or environmentally deprived by parents who may not have the biological or cultural resources to rear them successfully. The simple medical techniques that make eugenic sterilization possible emerged decades earlier than techniques used for more sophisticated biomedical interventions; we have gathered experience, formulated philosophies, worked through a process of maturing attitudes which we can now apply to newer interventions.

Eugenic sterilization seems like a small topic compared to its cousin, Population Control. Population Control envisions methods of temporary and reversible sterilization – licenses to procreate, drinking water additives which are not to be applied selectively to improve stock but are to be applied indiscriminately to reduce population density. Population Control has advocates who see it as imperative. "If our most treasured democratic institutions are to be preserved – and with all their faults, we know of none better – then birth control must be compulsory," says Robert Ardrey. "I do not accept the approach of eugenics. I do not believe we shall produce our Abraham Lincolns or our Albert Einsteins by favoring the rich against the poor, the high I.Q.'s against the mentally retarded. I trust the evolutionary process too implicitly, far more implicitly than I trust the judgment of men as to what qualities are of genetic advantage" (1).

Eugenic sterilization is the enforced sterilization of individuals who would not wish to be sterilized; the state has decided that it is in the public interest that this woman not produce progeny. Many would agree with Justice Holmes that *three generations of idiots is enough* . . . (2). In some European countries it is not the woman who is surgically sterilized – by salpingectomy or hysterectomy – but a male sexual offender who is castrated, but our usual frame of reference is the curtailing of a woman's procreativity. We can improperly minimize the importance of the topic by saying that the procedure was never widespread and in recent years its use has been declining. Paul states that from the peak of the 1930's, when nearly 25,000 operations were performed in the decade for eugenic or other considerations (out of a sixty-one year cumulative total of 61,000), the rate of reported state sterilizations in recent years has been running close to 400 or less, with nearly half of these coming from one state, North Carolina (3). But we are dealing with a procedure which like capital punishment should not be debated in terms of numbers; the ethical and moral and philosophical considerations of My Lai or of capital punishment do not depend on the number of victims. Then too, the number of sterilizations now reported yearly possibly do not include a number of *voluntary eugenic sterilizations,* those sterilizations which are consented to by the woman who is pressured by an institution, from which she will be released if she submits to the procedure but where she will stay if she will not; this is the practice of segregation of sexually promiscuous women, especially if they are retarded, during child-bearing years for which Morton Birnbaum has coined the telling term, *institutional sterilization*(4).

Whatever the figures, we want to consider our attitude towards the practice; we want to see this in relation to frequent proposals that the practice be extended to new areas, from the retarded to the recipient of welfare; and we want to see the problem in relation to other biomedical interventions which have related moral-philosophical-ethical considerations.

Medicine has the ability to intervene so as to confer life; it can prevent birth by its interventions; it also can confer death by its interventions. A constellation of related topics raises the question of the propriety of the medical intervention, the safeguards that should be interposed, the complicated moral-philosophical-ethical, ecological, and practical problems that the fact that we can intervene forces us to face.

Abortion is the most prominent of these topics. Euthanasia is the most frightening. Organ transplantation and the rights of donors are some of the newest in this constellation, along with the similar question of the prolongation of life by medical means and the criteria for the decision to stop prolongation.

In this category of topics the medical intervention will terminate the life process. It will terminate the life of the embryo or foetus in the case of abortion; it will terminate the life of the aged and the ill in the case of euthanasia; it will terminate the *life* of the potential heart donor (if his state of being, a state characterized by brain death but not by other characteristics of death, is in reality, life), and it also raises the possibility of future risk for donors who give up one of their kidneys, so in this latter situation we have the complicated problem of the information that is required to make an informed consent truly informed (5).

Eugenic sterilization is like euthanasia, organ transplants, the decision not to prolong life, and abortion in that it is a biomedical intervention dealing directly with the question of who shall live; it is unlike these because the intervention is to forestall creation, to serve as a permanent contraceptive, rather than to terminate a life within being.

Any sterilization does affect the life of the sterilized, but not in the same way that abortion affects the foetus or euthanasia the ill; assuming the operation goes well, the life or lives affected are the potential lives of children not yet conceived rather than actual lives. So sterilization resembles contraception, and since some contraceptives are probably in reality abortifacients (the expulsion of the fertilized ovum is presumably the method that the prostaglandins *prevent* conception), sterilization may have fewer moral-philosophical-ethical complexities than a number of other biomedical interventions including some methods of *contraception.*

But although sterilization itself may represent a widely accepted procedure with comparatively few problematical considerations, eugenic sterilization represents a narrowly applied procedure – we have said the annual rate now is less than four hundred – with the same kind of serious considerations raised by the other four interventions. The fact that it is an enforced procedure raises many of the same issues as heart donations, euthanasia and other major interventions.

Sterilization procedures, male or female, are classified as eugenic, therapeutic, or sterilizations of convenience. Sterilizations of convenience are performed on eager subjects who believe that this will be the most certain or the most satisfactory method of birth control. Therapeutic sterilizations are performed on willing subjects – the patient must give consent to the procedure – because medical opinion indicates that further pregnancies may endanger the life of the mother or be harmful to her health. In contrast to these voluntary procedures, eugenic sterilizations are performed on subjects who are presumably unwilling to have the procedure but who are required to have the procedure because the state believes they would not be fit parents or would have defective children (6).

Compulsory surgery smacks of the worst kind of authoritarianism. It is difficult to reconcile with Anglo-Saxon ideas of the preservation of the rights of the individual, and it has connotations of the Hitlerian ideas of fostering of racial purity through sacrifice for the state. For these and other less specific reasons – our willingness in recent times to tolerate greater degrees of departure from the average or the *normal*, our decreased stress on intelligence and ability as measures of worthwhileness – it has lost some of its popularity. Julius Paul notes that twenty-seven states still have sterilization laws on their books, that in recent years numerous legislative attempts have been made to include sterilization as part of a program of recommended punitive action for welfare recipients who are unfit parents or who have more than a stated number of illegitimate children (7).

All of the new punitive sterilization acts, which are noneugenic but which have a close relationship to our topic, have failed to pass, but we should nevertheless note, as Paul has, that states which have evidenced various degrees of interest in punitive sterilization in recent years are California, Delaware, Georgia, Illinois, Iowa, Louisiana, Maryland, Mississippi, North Carolina, and Virginia. He notes also that, although Pennsylvania has never had any compulsory, voluntary, or punitive sterilization (although there is evidence that 270 sterilizations were performed *without* benefit of law on the patients of the Elwyn State School between 1889 and 1931), sterilization was proposed by a Philadelphia judge to the mother of four illegitimate children who had murdered her fourth child, age four days; she had been found guilty of second degree murder and according to a statute since declared unconstitutional because men were not subject to the same indefinite term could have been held for life. Judge Raymond Pace Alexander pointed out that he could not compel a sterilization procedure but that he was suggesting voluntary sterilization *as the best solution in a situation of this sort* and that if the defendant was sterilized he would consider releasing her in three-and-a-half years. "This is a recommendation, not a compulsion . . . you ought to do it," said the judge. Paul asks, "Does a defendant, under these circumstances (and mentally retarded in this case) *fully* understand not only the nature of the offer, but the consequences (and especially the *long-term* ones) of such surgical action" (8)?

Another kind of pressure to make eugenic sterilization *voluntary* has been proposed by Nobel laureate physicist William Shockley of Stanford University at the 1971 annual convention of the American Psychological Association. "At a bonus rate of $1,000 for each point below one hundred I.Q., $30,000 put in trust for a seventy I.Q. moron of twenty child potential might return $250,000 to taxpayers in reduced cost of mental retardation care." Dr. Shockley's plan to give a bonus to low intelligence parents who submit to voluntary sterilization was not supported by others at the meeting; he was called a *paranoid* and *facist* and his basis for evaluating findings from an Army pre-induction test were attacked (9).

The *Philadelphia Inquirer* conducts phone polls on questions of current interest. To the question, "Should the U.S. Encourage Sterilization Among Low IQ Groups?", phoned unsolicited answers ran 69.2 percent in favor of such encouragement and 30.8 percent opposed (10). The figures have no significance because people who phone in opinions presumably represent a skewed population, and the results of other polls in the series show no consistent relationship to more scientific polls and to popular vote results on other issues but the individual comments do indicate current attitudes.

Sample *yes* comments included:

"Everyone on welfare should be sterilized."

"It should be encouraged among all groups."

"There are too many illegitimate, unwanted children."

"Survival of the fittest . . . "

"Any woman who has an illegitimate child should automatically be made sterile."

"We have enough idiots walking the streets now."

"Specialists have been advocating this for years, but the church won't go along with them."

"People with low IQ's shouldn't have more than one child."

"Everyone should have it done after the third baby."

Some of the *no* comments were:

"You can't legislate morality."

"It's a diabolical thought."

"Encourage education and you won't have low IQ groups."

"Hitler had the same idea."

"It would make us a Godless nation; never."

"Sounds like mass genocide to me."

"Only the high IQ groups . . . "

"There's already too much dictatorship in this country."

"Whose IQ test would you use and who would play God?"

"Rich people cause more problems than poor people."

"It would be cold-blooded murder."

Sterilizations of male defectives were carried out without authority of law in the United States long before our first legislative experience in 1897, when a bill to authorize the procedure was defeated in the Michigan legislature. A Pennsylvania bill passed the legislature in 1905 but was vetoed by Governor Pennypacker who thought its authorization *to perform such operation for the prevention of procreation as shall be decided safest and most effective* conferred too great powers. Said the veto message:

> This Bill has, what may be called with propriety, an attractive title. If idiocy could be prevented by an Act of the Assembly, we may be quite sure that such an act would have long been passed and approved in this State, and . . . in all civilized countries . . . The nature of the operation is not described, but it is such an operation as they shall decide to be safest and most effective. It is plain that the safest and most effective methods of preventing procreation would be cut the

> heads off the inmates, and such authority is given by the Bill to this staff of scientific experts . . . (11).

Indiana became, in 1907, the first state to enact a compulsory sterilization law. The Indiana law is a tribute to the ingenuity of Dr. Harry C. Sharpe of the Indiana State Reformatory who had developed, using his prison population as experimental subjects, a new method of sterilizing males, the vasectomy, to replace the older and cruder castration. During the same period that Dr. Sharpe was perfecting his technique, in the last decade of the nineteenth century, the now standard method of sterilizing the female, the salpingectomy, was developed in France.

The enforced sterilization laws of all those jurisdictions which passed them prior to 1925 were declared unconstitutional, but in 1925 statutes in Michigan and Virginia were upheld by the highest courts of those states, and the United States Supreme Court, in the famous case of *Buck* vs. *Bell* in 1927, upheld the Virginia law (12).

Buck vs. *Bell,* the 1927 Supreme Court case which upheld the right of the state to order sterilizations of unwilling women, is so frequently quoted and cited that its facts are worth retelling.

Carrie Buck was a feeble-minded white patient at the Virginia State Colony of Epileptics and Feeble Minded, the daughter of a feeble-minded mother in the same institution and the mother of an illegitimate feeble-minded child. A Virginia law stated that the health of the patient and the welfare of society might be promoted in certain cases by the sterilization of mental defectives. The procedure was set forth in the law: whenever the superintendent of a state institution was of the opinion that it was in the best interests of the patient and of society that a patient *afflicted with hereditary forms of insanity and imbecility* be sterilized, he could petition a special board of directors of the institution, a hearing would be held in the institution, and the board could make the determination; in order to have an effective right to appeal, all of the evidence was required to be reduced to writing. The patient or her representative could appeal to the Circuit Court of the County or take a further appeal to the Supreme Court of Appeals. The Holmes decision in the case is famous for its finding and its phraseology.

Justice Oliver Wendell Holmes, the Yankee from Olympus who had never forgotten his days as a soldier in the Civil War, and who carried his patriotism to the point of making the United States the beneficiary of his will, believed that the sacrifice of an individual forced to give up his opportunity for procreation and to submit to a surgical invasion of his scrotum (or her abdomen) was no greater than the sacrifices that the government requires of its soldiers during war, also for the common good. He stated that the case law upholding the right of states to enforce compulsory vaccination law against smallpox by analogy could apply to those who would not be *vaccinated* against another communicable disease, pregnancy (13).

The Holmes decision was based on the assumption that the defects were hereditary in the absence of any scientific evidence challenging this theory, and on the precedent of the Massachusetts compulsory vaccination law (which, however, provided that those who *should deem it important that vaccination should not be performed* had merely to pay a fine of five dollars to escape the procedure).

It has been pointed out that Justice Holmes may have been napping on Olympus when he used the Massachusetts vaccination statute's constitutionality as a precedent because the alternative fine was not comparable with enforced submission to sterilization and because the scientific findings of the efficacy of vaccination carried much more authority than the dubious claim that mental disorders are largely hereditary and that sterilization could lead to a great decrease in retardation.

The Holmes decision is the source of ringing quotations. "Three generations of imbeciles are enough," and "The principle that sustains compulsory vaccination is broad enough to cover cutting the Fallopian tubes." Said Holmes:

> We have seen more than once that the public welfare may call upon the best citizens for their lives. It would be strange if it could not call upon those who already sap the strength of the state for these lesser sacrifices, often not felt to be such by those concerned, in order to prevent our being swamped with incompetence. It is better for all the world if, instead of waiting to execute degenerate offspring for crime, or to let them starve for their imbecility, society can prevent those who are manifestly unfit from continuing their kind.

Twenty-seven states now have so-called involuntary sterilization laws – in all but six of these the procedure is limited to the residents of specifically-named institutions for female retarded or to institutionalized patients – but in several of these states the law contains the possibly contradictory statement that the consent of the subject must be obtained. The laws apply in various jurisdictions to one or more of the mentally deficient, the mentally ill, the epileptic, *hereditary criminals,* sex offenders, and in a few states even to syphilitics.

When attempts are made to extend this practice into new areas, or when eugenic sterilization is ordered without the authority of statutes specifically providing for the procedure, civil liberties are threatened in a dramatic way.

In 1966 Nancy Hernandez, a twenty-one year old Chicano mother of an illegitimate child was ordered by a Santa Barbara judge to serve six months for a marijuana offense but was offered probation if she agreed to sterilization. She was imprisoned, after agreeing to the operation, when she changed her mind and refused the operation, but she was quickly freed on a writ of habeas corpus and the original order was vacated. Said Judge C. Douglas Smith of the California Supreme Court:

> The imposition of sterilization in the present case by the judge of the Municipal Court was in excess of his judicial power. There is no statute or other law which authorizes it in such a case as this. It constitutes a punishment beyond the power of the court to impose . . . Only the Superior Court is given power to order sterilization of a human being and then in very limited special cases (14).

An important 1971 case concerned an Ohio woman who was sterilized by order of a judge in the absence of a statute authorizing the procedure. She was held entitled to bring suit against the judge in spite of the doctrine of judicial immunity, the court holding that the offending judge would have been immune from suit if he had merely exceeded his authority but in this situation was not immune since he had acted in the absence of any authority (15).

These and similar cases in recent years have led to the reconsideration of eugenic sterilization in two recent books – Meyers', *The Human Body and the Law,* which compares and contrasts the American attitude on this with those of nine other nations (16) and Kittrie's, *The Right to be Different,* in which sterilization is considered as a drastic therapy with some of the same implications as forcible electroshock therapy and lobotomy (17). The American Bar Foundation study, *The Mentally Disabled and the Law,* recommends that statutes authorizing involuntary sterilization should be repealed and *voluntary* sterilizations by substituted consent be carefully scrutinized; it recommends that as applied to the mentally disabled "no statute shall authorize sterilization requested as the result of a condition imposed for probation, or discharge from, or entrance into, an institution, and that all statutes shall require an inquiry with respect to such possibilities (18).

In recent years intrauterine contraceptive devices and depot hormonal injections have been advocated as alternatives to surgical procedures. The low cost and ease of insertion of intrauterine devices may make them preferable to the salpingectomy, but what is easily inserted can also be withdrawn; long-term injections are also not practicable at this stage of the development of the art. In either case, the questions could be raised of possible duress and the validity of the consent of the defective.

Just as the topic of heart transplants, affecting only a few hundred donors and donees, raises much material for thought and discussion, so the topic of eugenic sterilization raises interesting questions for our considerations. To what extent is the body a private area to be safequarded from invasion? To what extent does retardation give the state the power to invade? If a girl is institutionalized for low intelligence and promiscuity and told she can be released from the institution if she has her tubes tied, is her consent to the procedure voluntary? And if she is retarded how do we know she understands the long-range effects of giving up her ability to procreate?

Does low intelligence equate with poor abilities to mother? Do those states which encourage eugenic sterilization thereby discriminate against blacks or other racial minorities? Does eugenic sterilization, like population planning, have the genocidal potentialities that some blacks fear?

Although we are dealing nationwide with only a few hundred

retarded cases yearly, the questions that must be answered before we can feel comfortable in our decision to push for, or to oppose, eugenic sterilization go so deep that in the process of finding answers we have to examine our philosophy of life, our feelings about the responsibility of the individual, our feelings about the power of the state.

All topics pursued far enough come to the point where well-intentioned men find their differences irreconcilable. Then they have the job of finding new ways of living together in spite of ideological differences that are profound. Perhaps eugenic sterilization is a topic where these gulfs and rifts are particularly easy to see. Perhaps, (like the suggestion that the reaction to Dostoevsky provides an ink blot test that reveals the perceptive sets of the reader) eugenic sterilization is a topic which mobilizes amorphous reactions, forces them to be clarified, makes attitudes of the individual more apparent to himself and also reveals him more clearly to others.

Abraham Goldstein, the Dean of the Yale Law School who writes brilliantly on legal psychiatric topics, has discussed how practical decisions lead inevitably to moral issues and how political questions then become complicated with a differentiation of short run and long term solutions. Although he uses the environment and overpopulation as his examples, we find the same moral and political questions raised when we consider eugenic control:

> . . . In area after area, we find that the issues raised by the environmentalists are the merest beginning and lead rapidly on to moral positions which are deeply held, suspicions which are keenly felt, and assessments which must be made as to how society might best deploy inevitably limited resources.
>
> The question of overpopulation illustrates the difficulties very well. it is widely believed that anything we do about pollution, housing, welfare, or transportation is the merest of palliatives as long as our population continues to increase at the present rate. Yet the moment the call to action sounds, we find ourselves involved in a raging debate on the right to live and the right to die. Many people, out of the deepest religious conviction, find it abhorrent to think of birth control or abortion or euthanasia as a part, either of social planning or of what individuals may do in exercising free choice. Others, particularly blacks, are wary of birth control because they see it as a

> subtle effort to limit their numbers and thus prevent them from achieving political power, denying them what was achieved by Catholics in many communities in the past.
>
> What starts out, therefore, as the most basic target of the new effort is translated almost at once into a political question, with interest groups on one side or the other and practical judgments made as to whether it really is 'worth stirring up all this fuss.' Obviously, it is well worth stirring up a fuss because the future of society may depend upon it. But the appetite for dealing with the problem in such apocalyptic terms declines as the political component increases. The long run is sacrificed to the short run because we are not confident enough of our purpose (19).

The most recent biomedical intervention to sponsor controversy is cloning, a procedure not yet practicable but seen by many authorities as a possibility for the future. Cloning is the mirror image of eugenic sterilization; it asks for the increased fertility of those who are presumably eugenically superior. Gaylin has described the potentialities (20):

> Cloning is the production of genetically identical copies of an individual organism. Just as one can take hundreds of cuttings from a specific plant (indeed, the word *klon* is the Greek work for *twig* or *slip*), each of which can then develop into a mature plant – genetic replicas of the parent – it is now possible to clone animals. The possibility of human cloning seems to produce in non-scientists more titillation than terror or awe – perhaps because it is usually visualized as *a garden of Raquel Welches,* blooming by the hundreds, genetically identical from nipples to finger nails.

Whether our concern is who should or who should not reproduce, we have to ask many of the same questions, particularly questions on the state of our knowledge, on the determination of the people who make the determinations, on the effect of alteration of the conditions under which human life has evolved.

Eugenic sterilization is a topic which does not go wide but does go deep. It is, indeed, *worth stirring up all this fuss.*

REFERENCES

1. Ardrey, Robert: Birth in the wilds: II. New York Times, Sept. 28, 1971, p. 37.

2. Buck v. Bell, 274 U.S. 200 (1927), 207.
3. Paul, Julius: The return of punitive sterilization proposals: Current attacks on illegitimacy and the AFDC program. Law and Society Review 3 (1): pp. 77-106, Aug., 1968, p. 78.
4. Birnbaum, Morton: Eugenic sterilization. JAMA 175: 951, 1961.
5. The obverse of informed consent also raises complicated problems. For many years and in some parts of the country currently, competent individuals who understand a procedure completely are not allowed to submit to the procedure because the physician, without legal basis for his refusal, feels the procedure might be unwise. One example is the surgical sterilization – not a eugenic but a voluntary sterilization – of single people. Another is the case of a Wisconsin physician who has been trying for eight years to give a blind person one of his eyes; the offer has been consistently refused; Atlanta Journal and Constitution, Sept. 26, 1971, p. 11-B.
6. Robitscher, Jonas: Pursuit of Agreement: Psychiatry and the Law. Philadelphia, Lippincott, 1966, pp. 79 ff.
7. See footnote 3, supra.
8. Idem., p. 96 quoting Philadelphia Inquirer, June 1, 1966, p. 11 and New York Times, June 2, 1966, p. 37.
9. Continuing the I.Q. controversy. Science News Letter 100: 190, 1971.
10. The public speaks. Philadelphia Inquirer, Sept. 23, 1971, p. 33.
11. Challener, William A., Jr.: The law of sexual sterilization in Pennsylvania. Dickenson Law Review 57:298, 1952.
12. See footnote 2, supra.
13. See footnote 6, supra, p. 81.
14. In the Matter of Hernandez, No. 76757 Santa Barbara Superior Court, June 8, 1966; see Shearer, L.: Should this woman be allowed to have more children? Parade, Aug. 7, 1966, pp. 4-5.
15. Wade v. Bethesda Hospital, Civil Action 70-225 (D.C., Ohio, Sept. 8, 1971).
16. Meyers, David W.: The Human Body and the Law: A Medico-Legal Study. Chicago, Aldine, 1970, pp. 26-47.
17. Kittrie, Nicholas N.: The Right to be Different. Baltimore, Johns Hopkins, 1971, pp. 297-339.
18. The Mentally Disabled and the Law, Revised Edition. An American Bar Foundation Study (eds. Brakel, S.J. and Rock, R.S.), Chicago, University of Chicago, 1971, pp. 207-225.
19. Goldstein, Abraham: Wall Street Journal, July 16, 1970, p. 7.
20. Gaylin, Willard: We have the awful knowledge to make exact copies of human beings. The New York Times Magazine, March 5, 1972. p. 12.

Chapter II

SOME THOUGHTS ON EUGENIC STERILIZATION

GEORGE TARJAN

I AM a child psychiatrist. This background makes me fully aware of and appreciative of, the importance of good parental and adequate child rearing. I am convinced that, on the whole, without the benefits of good parents and adequate parent-child relationships, most children start life with several strikes against them. On the other hand, also as a child psychiatrist, I have observed time and again the resiliency and the recuperative power of children. I noted often that many children grow into adults with greater competence in intellectual functioning than the prognostications of professionals would have assigned to them earlier.

As a professional individual whose primary interest has been mental retardation for some twenty-three years, I have learned a few relevant things about this condition. The label of mild mental retardation – and this is the degree of retardation with which we are generally concerned when we talk about sterilization, particularly eugenic sterilization – depends as much on external circumstances and the predilections of examiners as on the symptoms of the patients. I have found that from among a group of individuals who might qualify for the diagnosis of mild mental retardation, only certain individuals are singled out for actual labeling. The factors which lead to the diagnosis are often environmental rather than internal to the patient's performance.

NOTE: Supported in part by Federal Grants: HD-08667; MH-10473; MCH-927; and by the Stanley W. Wright Memorial Fund.

I will not belabor the definition or the description of mental retardation *per se,* but will make a few further points about mild mental retardation. The diagnosis usually is not assigned to an individual prior to school age and it generally disappears after the person leaves school. Hence, the diagnosis is not only age specific but also age limited. Further, the mildly retarded are much more capable of adequate performance in daily life than many clinicians believe. A large segment of this group was, in fact, recently described as *the six hour retarded* referring to the usual failures in school as contrasted with relative successes elsewhere (1). The mildly retarded are capable of making rather important decisions pertaining to practical, moral, and ethical matters. Finally, they generally show a greater degree of self-control and self-governance than many experts assume. As a consequence, I have long believed that we professionals are inclined to overprotect the retarded once we make such a diagnosis. I would favor permitting them greater independence, including the right to make some inevitable mistakes.

Soon after I assumed responsibility, nearly twenty-five years ago for Pacific State Hospital, Pomona, California, an institution for the mentally retarded, I learned that both the availability and the performance of voluntary or involuntary sterilization depended foremost upon the fact of admission of the individual into the state hospital. Admission, on the other hand, just as diagnosis, depended upon external circumstances as much as on the symptomatology of the patient. In those days state hospital stay was seen as non-time limited for the mildly retarded, whereas today very few such patients even enter these hospitals, and those who are admitted leave after a relatively brief inpatient phase.

During my period at Pacific I found that the near elimination of sterilization did not produce major changes in the mode of living or behavior of the patients, nor did it result in the calamities some people predicted. Two events occurred almost concurrently. One was the rapid decrease in the number of sterilizations, and the other was a major increase in patient freedom. With the latter came greater opportunities for unsupervised interaction among patients of the two sexes. Many people, particularly those living around the hospital, thought that the increments in patient freedom and the decrements in sterilizations would certainly produce an enormous increase in illegitimate conceptions and births. It was even thought that an observable increase in the frequency of mental retardation might result. The facts did not bear out these predictions. The conception rate was so low, essentially nil, that it could not satisfy the requirements of a study of the frequency of mental retardation. However, on the basis of clinical impressions, the illegitimate pregnancy rate of Pacific's patients could have given pride to any neighboring college or even high school.

Some of us were intrigued by this outcome, which was rather unexpected by many people. Fortunately, our group included an anthropologist, Dr. Robert Edgerton, who wrote a paper based on his observations (2). He noted that dating relationships were aggressively exclusive, that everyone who was dating among the patients was going steady, that fidelity was highly valued, was demanded and received from partners, that rules of proper conduct were defined to include a decidedly puritanical sexual life, that sexual ethic was considerably more delicate than that generally in effect outside the hospital, that control of sexual impulses was remarkably outstanding, and that the retarded learned to accept the rules of others as well as their own as the right rules and responded punitively to breaches of these rules. It is difficult in the face of these observations to be an enthusiastic supporter of sterilization as a eugenic tool in the field of mental retardation.

There are, however, critical issues to be examined including certain aspects of sterilization, the question of volunteering for the procedure, and a few points about eugenics. I do not believe that sterilization needs definition. It obviously has to be clearly separated from castration, though occasionally historically the two have been intermixed. However, even as of today, when we speak of sterilization in the context of eugenics, we are talking about a procedure that is intended to be permanent.

As a consequence, it is imperative to give some thought to the implications of this permanence. Again, some of my personal experiences are relevant. At Pacific I was frequently confronted with letters written to me by physicians inquiring about the nature of surgery that was performed, particularly on women patients, who were seen later by them, as married adults. Usually, the patients themselves sought surgical repair. The physicians felt that whatever the indications for sterilization might have been at one time, they were no longer valid. It was my custom to explain the legal basis for the operation and then describe the procedure performed. Further replies from the doctors focused on the fact that our former patients could not be easily identified as being mentally retarded during their adulthood. They functioned like many other individuals and frequently proved to be adequate mothers to their husbands' children by prior marriages.

On one occasion when I spoke about the above observations and recounted the success in child rearing noted by Harold M. Skeels in retarded adolescent and young adult women (3), a press reporter quoted me as having said essentially that *mentally retarded girls make the best mothers.* Obviously, I said nothing of the sort, but I did state that in my judgment a high IQ is not the most essential requirement of good mothering and that many women with a relatively low measured level of intelligence can function well in this role. All these facts compel me to emphasize that the permanent nature of sterilization must be given careful consideration.

Another issue to be examined is voluntary sterilization as contrasted with involuntary procedures. The practices which had been used in some of the states, at least as far as mentally retarded individuals were concerned, were basically nonvoluntary. They were either legally mandated or permitted. The aims were clear and can be exemplified by the historical practice of one state which did not have legal provisions for sterilization but maintained an *institution for mentally defective women of childbearing age.*

The question of volunteering raises further issues, particularly in the mentally retarded. People often assume that persons whose measured level of intelligence is below a specified cutoff point, have sexual proclivities or inclinations which are entirely different from those noted in people who happen to have an IQ of some points higher. Furthermore, it is also assumed that control of sexual impulses has a straightline relationship to IQ. I believe that these assumptions are rooted in exaggerated emotions as much as in hard facts.

It is also important to realize that concern with sexual activities and therefore preoccupation with sterilization, both in the parents of the retarded and in the general public, usually arises around the puberty of the child under consideration. If we are talking about voluntary sterilization, we must note that the issues are usually decided when the person in question is an adolescent and therefore not yet of age. Children in general are not considered to be mature enough to authorize even such surgery as tonsillectomy, which obviously is of lesser consequence than sterilization. Unfortunately, retarded children and adolescents are considered even less competent than those of average intelligence. In many state laws the definition of retardation includes some reference to the person's incapacity to manage his own affairs with ordinary prudence. It is doubtful that their consent could be considered as a truly informed one.

Retarded persons, particularly children, have also been noted to be highly suggestible. I have no doubt that an authoritative figure could readily influence a retarded adolescent to follow any suggested course of action, including sterilization. These facts cast further doubt on the propriety of taking the consent of an adolescent retarded person at face value. I would probably feel differently if the issue generally arose during adulthood because I think that retarded persons have the capability to make many decisions pertaining to their own welfare.

Under these circumstances, who should have the clear right of giving an informed consent for a retarded person who is not of age? The obvious answer might seem to be the parents. I hope those more knowledgeable in the area of the law will further debate this question. I only want to express some doubts of my own. We know that many mentally retarded adolescents become quite well adjusted adults. Yet the consequences of the surgery remain essentially permanent. Therefore, should parents – based on symptoms which manifest themselves during adolescence but may disappear later – be permitted to deprive their children of the opportunity to reproduce?

It is also often assumed that mentally retarded individuals do not have the emotional capacity to respond to the consequences of sterilization. These assumptions are not congruent with the findings of Edgerton (4), who studied retarded adults whose main life goal was to shed the label of retardation and to pass as normal persons. Some of them saw their past surgery as a permanent *brand,* constantly reminding them of the diagnosis which was attached to them during adolescence.

This, then, brings me to some thoughts on eugenics. First, a definition which I borrowed from an early leader in the field, Charles Davenport. He stated, "It is the science of the improvement of the human race by better breeding" (5). If one reads the eugenic literature one finds that it was the eugenicist's hope to create a society in which each child would be born with vigorous health and able mind through the encouragement of the propagation of those with desirable traits and through the restriction of propagation of those with undesirable traits. This approach, as you can see, included a balanced portfolio. Most of eugenics and most eugenicists, however, were committed to only one side of the coin. It was the restriction of propagation of those who were seen, correctly or incorrectly, by certain arbiters of social judgments as not being worthy of the privilege of reproduction. Much less was said about the encouragement of reproduction of individuals with desirable traits, and even the few voices in this direction almost disappeared under the pressures of concerns with population explosion.

Second, I am concerned with the logic that equates for the purposes of reproductive rights, *goodness* and *badness* with *high* and *low* IQ. I am not convinced that a relatively high IQ assures the concurrent presence of many other eugenically desirable traits, or that a low IQ necessarily means the coexistence of many undesirable ones. In my experience, individuals with the same IQ, high or low, are quite variable in other personality characteristics. One must remember that intelligence tests were constructed to prognosticate success or failure in the ordinary classroom. They serve well in this respect. Though the IQ also predicts performance in other selected life situations, I doubt that it represents a good single scale upon which an individual's contribution to human evolution can be assessed.

It is important to emphasize the fact that the IQ is expressed on a numerical scale on which some cutoff point is needed for practical decision making. Sterilization is not a halfway measure, it is either performed or not performed. If the decision of surgery is based on the IQ, some cutoff point must be used for individuals as well as groups. However, the variability of the results obtained on the same individual at different times exposes any decision to the caprice of the limitation in reliability. Further, even if this problem of reliability were to be solved, who could categorically decide that – because of a few extra IQ points – a person is more desirable from a eugenic point of view than his somewhat less endowed counterpart?

So much about the role of measured level of intelligence in eugenics. Someone might argue at this point that a scale, or a set of scales, could be devised which would measure the individual's potential contributions to mankind's evolution. First, I doubt that this goal is realistic, and second, I contend that the same problems of reliability would still hold.

I must also raise the question as to whether our increased knowledge in genetics, biological engineering and euphenics rendered many of our historical notions on eugenics anachronistic (6). At one time it was customary and reasonable to speak about the *nature-nurture controversy.* Scientific evidence of those days suggested that innate characteristics were not readily influenced by external forces. Today, we have sufficient evidence to realize that the two seemingly independent realms constantly interact and affect one another. The dietary treatment of phenylketonuria is a simple and straightforward example focusing on our ability to manipulate, environmentally at least, the phenotypic expressions of genetic traits. It is more proper, therefore, to speak about *nature-nurture interaction.* In this context, many traditional eugenic concepts may well be based on a one-sided approach to a two-sided issue.

I have raised many questions for consideration. The problems of retardation, in particular, and that of eugenics, in general, are relatively small components of a much larger dilemma – namely, that of the explosive growth of the human population. This is an overriding issue. I hope that mankind will have the moral courage, the strong determination, and the ethical competence to solve this great problem. Mental retardation and eugenics, I believe, will then fall into perspective.

REFERENCES

1. The Six-Hour Retarded Child, A Report on a Conference on Problems of Education of Children in the Inner City - Aug. 10-12, 1969. Airlie House, Warrentown, Va. President's Committee on Mental Retardation, Washington, D. C. 20201.
2. Edgerton, Robert B. and Dingman, Harvey F.: Good Reasons for Bad Supervision: "Dating" in a Hospital for the Mentally Retarded. Psychiatric Quarterly Supplement, Part 2, 38: 221-233, 1964.
3. Skeels, Harold M.: Adult Status of Children with Contrasting Early Life Experiences. Monographs of the Society for Research in Child Development, Serial No. 105, 31, (3), 1966.
4. Edgerton, Robert B.: The Cloak of Competence – Stigma in the Lives of tne Mentally Retarded. Berkeley and Los Angeles, University of California Press, 1967.
5. Haller, Mark H.: Eugenics – Hereditarian Attitudes in American Thought. New Brunswick, Rutgers University Press, 1963, p. 3.
6. Lederberg, Joshua: Molecular Biology, Eugenics and Euphenics. Nature. 198:428-429, May 4, 1963.

Chapter III

STATE EUGENIC STERILIZATION HISTORY: A BRIEF OVERVIEW

JULIUS PAUL

STATE compulsory eugenic sterilization found its way into American law for the first time in Indiana in 1907. Until the famous case of *Buck vs. Bell* in 1927, its constitutional status was still in doubt, but Mr. Justice Holmes and the United States Supreme Court left no doubts about either its constitutionality or its vitality as a proper exercise of state police powers. State laws and sterilization operations peaked in the 1930's. The war years brought a slowdown, and after World War II, a vastly larger and more articulate mental health movement, aided in part by Federal funding, took a new look at the origins and treatment of mental illness and mental retardation and the utility of eugenic sterilization laws. The result was a precipitous drop in operations, either because of a refusal to enforce the laws, or a shift from compulsory procedures, or even outright repeal of eugenic sterilization laws (Kansas and North Dakota, 1965; Nebraska, 1969; Montana, 1969; Georgia, 1970).

By the mid 1960's, the annual cumulative total for the country was under four hundred operations, and more than half of these were in one state, North Carolina. Even North Carolina, with its additional voluntary sterilization law, and increased use of birth control and family planning programs, saw a reduction in board-ordered sterilizations. If North Carolina is running close to 150 operations per year, and the other twenty-five states with sterilization laws are doing a total of 150 or less, then eugenic

NOTE: Materials for this paper were taken from the author's unpublished manuscript, "... Three Generations of Imbeciles Are Enough ..." –State Sterilization Laws in American Thought and Practice (Washington, D. C.: Walter Reed Army Institute of Research, 1966).

sterilization in America has reached a low point, and it would appear that legislative efforts to revamp some of the present laws, especially where voluntary measures have replaced compulsory procedures (Connecticut, 1965, 1969; Georgia, 1970; and Montana, 1969) may mean that short of outright repeal, state eugenic sterilization will remain a very selective device at best.

Two other important trends in recent years deserve attention. The first took place in the late 1950's and 1960's, and was an effort to revive state sterilization not for eugenic purposes, but for *punitive* purposes, aimed mainly at mothers of two or more illegitimate children receiving AFDC or other welfare assistance. The many ill-fated efforts to enact such laws stem from a widespread effort to reexamine welfare laws and welfare costs, and from an often erroneous linkage in the public (and legislative) mind between welfare and illegitimacy. Although these measures have failed to pass, the battle against rising welfare rolls still persists, and punitive measures that employ sterilization may still see the light of day.

The other more recent development is the rapid – almost overnight – rise in the use of private, voluntary sterilization as a form of permanent (though sometimes reversible) birth control, and especially the use of the male operation, vasectomy. Four states (Georgia, North Carolina, Oregon, and Virginia) have enacted voluntary sterilization laws that are totally separate from any existing eugenic sterilization laws. A recent study of 1970 figures says that 750,000 Americans of both sexes underwent contraceptive sterilization in that year, and three-fourths of these were men. Only a few years ago, the comparable figure was 100,000, and the majority of those were women (1). Hence, voluntary private sterilization – and particularly vasectomy – has jumped enormously as zero population growth (ZPG) efforts and the contraceptive revolution have become a staple of American life in the 1970's.

It would seem that in the light of this upsurge in voluntary sterilizations and the increasing public (though not always medical) acceptance of voluntary sterilization, that compulsory state sterilizations would remain low and in some states with compulsory laws, the law would continue to be ignored. Eugenic sterilization is, for all intents and purposes, a dead issue. But is it really? It is conceivable that as birth control and family planning programs increase (particularly with the vastly expanded Federal programs under Johnson and Nixon) that sterilization might be used as one of several options, especially where multiparity is concerned, and the old game of *quid pro quo* (sterilization for welfare benefits) might appear again. The poor, and particularly blacks and other minority groups, are especially vulnerable to this kind of possibility. While the veneer would be *economic* (or ZPG), the underlying motivation would be otherwise. Hence, sterilization as a means for limiting population might fall on particular groups, and as in the earlier days of the extreme hereditarians, limit the expansion of those particular groups that legislatures find socially undesirable (or *cacogenic,* to the use the language of earlier days).

So long as voluntary sterilization is kept separate from public welfare programs, the eugenic element will probably be absent. But once government enters, there will be a strong temptation by some legislators (and governors) to let the chips fall where they may. Even under present voluntary sterilization laws, eugenic considerations can be allowed. Conceivably, voluntary laws with strict *informed consent* procedures could completely replace the compulsory sterilization laws. The difficult problem is how to apply consent procedures to mentally retarded persons who are not considered *competent* to make this decision for themselves. States such as Montana and Vermont have demonstrated that vigilant consent procedures can be instituted where eugenic considerations are concerned.

State sterilization history in America has come nearly full circle, from its extremely compulsory and punitive beginnings, to a recent upsurge in private, voluntary operations. The politics of public welfare and the so-called war on poverty will determine whether or not the social Darwinian origins of eugenic sterilization come back to haunt us in disguised forms. It would seem tragic indeed if the lessons learned from the sixty-five year history of compulsory state eugenic sterilization in America, together with the findings of modern medicine, psychiatry, and genetics, are a total nullity, and the epitaph of this epoch in our history still remains: *Three generations of imbeciles are enough.*

A review of those sixty-five years, with special emphasis on Pennsylvania's innovative role in the history, should set forth these lessons.

Although Indiana's Dr. Harry Sharp can claim the first use of vasectomy in a public institution in America, Pennsylvania can claim two firsts in the history of sterilization in America: the first known sterilization (more accurately, castration) in America in a public institution, and the passage of the first state sterilization law. It is ironic that notwithstanding these two firsts, Pennsylvania has never had a sterilization law on its books, despite persistent efforts.

Pennsylvania's place in the history of state sterilization in America began in the 1880's with the famous Dr. Isaac Newton Kerlin, then superintendent of the Pennsylvania Training School for Feebleminded Children at Elwyn. It was out of Elwyn that two subsequent figures in the sterilization movement appeared, Dr. Martin W. Barr, who succeeded Kerlin as superintendent at Elwyn, and Dr. A. W. Wilmarth, as Assistant Superintendent under Kerlin who later went on to fame in Wisconsin eugenic sterilization history; Kerlin and Barr are the chief figures in the Pennsylvania story.

Kerlin reported his first sterilization (probably castration) at Elwyn in a speech he delivered as president of what is now known as the American Association on Mental Deficiency in 1892, declaring to the assemblage: "What [whose] state will be the first to legalize this procedure?" He claimed he did this operation in 1889, but there is considerable confusion over the dates of 1889 and 1892 (2).

The spark lit by Kerlin was largely carried by Barr, who not only continued to castrate at Elwyn without benefit of statute, but who was also instrumental in the preparation and passage of the first sterilization bill in the country in 1905. Although both houses in Pennsylvania approved the bill, it was vetoed by Governor Samuel W. Pennypacker, whose famous veto message has this to say about *An Act for the Prevention of Idiocy:*

> . . . The subject of the act is not the prevention of idiocy, but it is to provide that in every institution in the state, entrusted with the care of idiots and imbecile children, a neurologist, a surgeon and a

> physician shall be authorized to perform an operation upon the inmates *for the prevention of procreation.* What is the nature of the operation is not described but it is such an operation as they shall decide to be *safest and most effective.* It is plain that the safest and most effective method of preventing procreation would be to cut the heads off the inmates, and such authority is given by the bill to this staff of scientific experts. It is not probable that they would resort to this means for the prevention of procreation, but it is probable that they would endeavor to destroy some part of the human organism. If this plan is to be adopted, to make it effective it should be carried into operation in the world at large, and not in institutions where the inmates are watched by nurses, kept separate, and have all the care which is likely to render procreation there very rare, if not altogether impossible. To permit such an operation would be to inflict cruelty upon a helpless class in the community which the state has undertaken to protect. However skillfully performed, it would at times lead to peritonitis, blood poisoning, lockjaw and death (3).

EUGENICS AND THE EARLY BEGINNINGS OF STATE STERILIZATION LAWS

Even before the rediscovery of Mendel's theories of inheritance and the efforts to translate Galton's theories into eugenics legislation at the beginning of this century, sterilization (specifically castration) was regarded as the solution to a number of medical and social problems, including insanity, criminality, and the allegedly terrible consequences of masturbation. Dr. William Goodell, a pioneer American gynecologist, optimistically wrote:

> I am, indeed, not sure that in the progressive future it will not be deemed a measure of sound policy and of commendable statesmanship to stamp out insanity, by castrating all the insane men and spaying all the insane women (4).

In spite of the fact that castration was used in some states *without* benefit of law (principally by F. Hoyt Pilcher in Kansas and Isaac Newton Kerlin and Martin W. Barr in Pennsylvania), it was not a popular *solution.* But Harry C. Sharp's pioneer use of vasectomy on youthful inmates of the Indiana Reformatory (without benefit of law) in 1899-1907 did *catch on,* and state sterilization was here to stay.

The end of the nineteenth century, with its emphasis on social Darwinism, brought in its wake a number of important intellectual, medical, and social developments, and a combination of events catapulted eugenics into the public arena in America. These included August Weismann's attack on the theory of acquired characteristics, the rediscovery of Mendel's theories of inheritance, the spread of Galton's philosophy of eugenics from England to America, the growth of intelligence testing, especially the Binet-Simon test, the rash of family pedigree studies, and statistics on prison and mental hospital populations, together with the discovery of two simpler and safer surgical procedures (vasectomy and salpingectomy) that could sterilize without castrating.

Eugenics, in the words of the historian Richard Hofstader, "has proved to be the most enduring aspect of social Darwinism" (5). And the most enduring relic of the eugenics movement in America is the existence of state sterilization laws, which are still present in twenty-four of our states. The myth that America would be deluged by the *cacogenic* or socially inadequate, and particularly the feeble minded, resulted in the rapid passage of eugenic sterilization laws by sixteen states in the period of one decade, 1907-17.

However, the admonition of one writer, who said that we should "Castrate the criminal, cut off both ears close to his head and turn him loose to go where he will . . . " was not followed in the laws, since the mentally ill and mentally retarded were the focus of most of the laws, and where the laws were punitive in nature, they were either nullified by the courts or seldom enforced (6). The report of the Committee to Study and Report on the Best Practical Means of Cutting Off the Defective Germ-Plasm of the American Population came much closer to what the state legislators were trying to accomplish, when it said: "Society must look upon germ-plasm as belonging to society and not merely to the individual who carries it" (7). This idea found a more specific constitutional formulation in the United States Supreme Court's landmark decision in *Buck vs. Bell* in 1927 upholding the constitutionality of compulsory state eugenic sterilization laws. This decision gave a credibility to compulsory state sterilization that the scientific community in America was never able to provide. Armed with the cloak of *constitutionality,* state sterilization laws spread into other states and the decade of the 1930's say upwards to 25,000 sterilizations performed under these laws (out of a cumulative national total of approximately 64,000 in the years 1907-63).

Some of these laws, in addition to the usual categories of insanity, feeble-mindedness, idiocy, imbecility, and epilepsy (using the stated language of these laws), also included rapists, habitual criminals, persons with *criminal tendencies,* drunks, drug fiends, syphilitics, moral and sexual perverts, *moral degenerates,* and *other degenerate persons.* Landman reported that as many as thirty-four distinct classes of persons were eventually covered by our various compulsory state sterilization laws (8).

PENNSYLVANIA, WITHOUT LAW, CONTINUES ITS EFFORTS

The veto by Governor Pennypacker did not sound the death knell to sterilization (in fact) or to legislative efforts in Pennsylvania. For another twenty-five years, the battle was waged in the state legislature and at Elwyn.

Efforts to pass a bill followed in 1911 (died in committee in the Senate), 1913 (died in committee in the Senate and defeated on final passage in the House), 1915 (died in both houses in committee), 1917 (died in House committee), and 1919 (reported out of House committee but with negative recommendations). Finally, another bill did manage to pass in 1921, but was vetoed by Governor William C. Sproul on grounds that it was in violation of the equal protection requirements of the Fourteenth Amendment in not applying to persons at large (9).

From 1929 through 1951, the superintendent at Elwyn, Dr. E. Arthur Whitney, presented sterilization proposals to every biennial session of the state legislature, all of which failed. In an article written with the collaboration of Mary McD. Shick, Doctors Whitney and Shick reported a total of 270 sterilizations performed at Elwyn since Kerlin's first up to the year 1931, all without benefit of law (10). It is not clear if these were *voluntary* operations in the sense that some kind of *consent* was obtained, nor is it made clear how many of these operations were castrations on male and female inmates.

While Elwyn was perhaps exceptional, it is important to note that not all of the institutional authorities in Pennsylvania agreed with the program of Kerlin, Barr, and their successors. In a 1930 discussion at the 54th annual meeting of the American Association for the Study of the Feebleminded, Dr. Harvey M. Watkins, then superintendent at the Polk State School and Hospital, suggested a limited program of *selective* sterilization of selected morons and carefully selected high grade imbeciles who are eligible for parole:

> It would be well first to have it a *permissive law rather than a compulsory law.* Such a permissive law, plus a program of conservative and sane education on the part of the public and relatives, tried out over a period of years, *carefully analyzing those cases done, would point the way toward further improvements and recommendations as to statutory changes (11).*

This author's correspondence with Pennsylvania authorities produced the following reactions. The superintendent at the Elwyn School, Dr. Gerald R. Clark, reported that: "Sterilizations are no longer performed at Elwyn" (12). The superintendent of the Laurelton State School and Hospital, Dr. Bernard A. Newell, wrote that: "Sterilization operations are not permitted by Pennsylvania law" (13).

The Deputy Attorney General, Mr. Edgar R. Casper, did a brief study of the matter on his own and reported that the records at Elwyn were *inadequate* and Dr. Whitney was unavailable for comment because of prolonged illness. He also reported that Dr. John E. Davis, Commissioner of Mental Health for the state, had stated that "present policy prohibits any sterilization in any State School, hospital or institution for purely eugenic purposes" (14).

The most intelligent and detailed reply came from Dr. James E. McClelland, the superintendent at the Polk State School and Hospital. He points out that sterilization was never a part of the formal and officially recognized program at Polk *at any time in its history,* and he added, "It could not well be, for there was never any Pennsylvania statute either approving or regulating such matters." Superintendent McClelland's experience covered twenty-two years at Polk. After discussing the relative merits of the Whitney-Shick and Watkins papers on selective sterilization, Dr. McClelland made some subjective judgments of his own about the general question of sterilization by law:

> I have a few other thoughts on this matter which are not primarily medical, and would like to take this opportunity to set them down. This is an age of steady encroachment on the rights and personal liberties of individuals. What can be done to somebody because he is mentally retarded may sometime become legally permissible to another individual, not because he is mentally retarded, but because there is something else the matter with him. This process can go on somewhat in the matter that the sea nibbles at a hole in the dike. Since there are so many individuals in high places today who feel that they know better what is good for the ordinary citizen than he does himself, is it not possible conceivably, that an initial encroachment on the right of a particular kind of citizen to reproduce, if he can, might eventually lead, over a period of decades or centuries, to governmental decree establishing euthanasia at age 150. I know this sounds far-fetched, but I have observed and have read that the process by which governments gain control of the lives of their citizens often starts quite insidiously. To my mind, therefore, the Governor who vetoed the act in 1905, whoever he was, has my admiration (15).

But, notwithstanding the vetoes by Governors Pennypacker and Sproul and the repeated rejections by the Pennsylvania Legislature of various sterilization proposals, efforts continue, sometimes even the most extreme ones. Such an example was State Senator James Berger's 1961 bill providing for the emasculation (castration) of certain sex offenders. Highly vocal opposition to the bill from such groups as the American Civil Liberties Union Pennsylvania branch prevented passage of this proposal (16). A murder trial in Philadelphia in 1966 provided an unusual background for the introduction of the subject of sterilization in the judge's discussion of sentencing (17).

Needless to say, from Kerlin to Barr to Berger, Pennsylvania sterilization (or asexualization) history has been somewhat bizarre. Though there is little humor in the Pennsylvania sterilization story, perhaps the following anecdote about Governor Pennypacker might suffice to conclude the story. The Governor was invited to a newspaper dinner at the end of the legislative session, which was a kind of gridiron club affair of that day. When he began to speak, he was greeted with catcalls, whistles, and booing in the usual manner of these dinners. After some minutes of pandemonium he raised his arms for silence and then squeaked out in his funny high-pitched voice: "Gentlemen, gentlemen! You forget you owe me a vote of thanks. Didn't I veto the bill for the castration of idiots?" It was reported that this brought down the house and assured him a respectful hearing from there on (18).

It behooves those who would revive interest in the subject of eugenic sterilization in the 1970's to reassess this earlier history of sterilization in Pennsylvania together with a sober and searching look at the very limited enforcement of state sterilization laws in the country today (19).

DEVELOPMENTS SINCE THE END OF WORLD WAR II

Since the end of World War II, there has been a precipitous decline in annual state sterilizations reported, due mainly to changes in the philosophy and administration of state mental health laws. With the exception of North Carolina, most of the leading sterilization states have either stopped using their laws (through what might be called administrative *repeal*), repealed them (as in the case of Kansas, North Dakota, Nebraska, Georgia, and Puerto Rico), or greatly reduced the incidence of annual sterilizations. However, a new category of *fitness for parenthood* has emerged in the past fifteen years, reminiscent of the kind of punitive proposals that were aimed at various categories of criminals and the *cacogenic* in the late nineteenth and early twentieth century. But these new proposals are not aimed at the *cacogenic* persons that Landman found listed under thirty-four categories, but mothers of two or more illegitimate children receiving Aid to Families with Dependent Children (AFDC) or other public aid, or even both parents of illegitimate children. These abortive attempts to return to punitive sterilization as a kind of collateral attack on illegitimacy, public welfare, and at times a not-too-covert attack on the underprivileged classes of American society, are clothed in a new jargon *(population explosion; ecology; environmental health)* (20). At the moment when compulsory state sterilization is but a tiny fraction of its earlier enforcement, we have returned to the punitive arguments and with the same economic, racial, and *moral* overtones that were advanced in the days of the extreme hereditarians (21).

Sixty-five years of state eugenic sterilization experience in America should have taught us a few lessons. But history never really teaches men if they are still enslaved by the mythology of their predecessors.

Three generations of imbeciles are enough is no longer a living tribute to the wit or the wisdom of Mr. Justice Holmes, nor is the surgeon's scalpel the *solution* to our most pressing social and economic problems, nor have we as yet found the secret to producing the highest *quality* of population (22). The impact of modern science on government will make the problem of protecting individual rights and individual privacy all the more imperative in the years to come. Future generations of Americans may believe that one's germ-plasm belongs to society and not merely to the individual who carries it, but when and if this happens, we can only hope that those basic procedural protections that protect the solitary individual against the state – that require and maintain a standard of accountability, in short, the rule of law – are still part of the American dream and reality (23).

REFERENCES

1. The best known organization in the field, AVS (The Association for Voluntary Sterilization, Inc.) reported a total of 750,000 operations in 1970, with three out of four on men. A study and survey organization called Lea, Inc. of Ambler, Pa. reported that 750,000 vasectomies were performed in 1970. (New York Times, July 5, 1971, at p. 16). To show how fast-growing this phenomenon is, the executive director of AVS, John R. Rague, was quoted in the latter news item as saying that "probably three million Americans of childbearing age, both men and women, have now obtained sterilization." In an earlier article, "Sterilization is an answer for many," New York Times, January 18, 1971, at p. 24, the AVS was reported to have said that almost 2 million Americans of child-bearing age have had sterilizations "as their method of permanent contraception." In less than 6 months, the estimable AVS has revised its figures from 2 million to 3 million, and the figures for 1971 are expected to considerably top these figures. If and when voluntary sterilization peaks is anybody's guess.
2. A later superintendent at Elwyn, Dr. E. Arthur Whitney, says that Kerlin did the first sterilization at Elwyn in 1889 with the consent of the parents. "Some stalwarts of the past," 57 Am. J. Ment. Def. 345 (1953)

p. 349. Barr said that the first two castration operations were performed in 1892 in his article, "Some notes on asexualization; with a report on eighteen cases," 51 J. Nerv. and Mental Diseases 231 (1920) p. 233. The confusion is probably due to the fact that Kerlin did not report his 1889 operations until the 1892 meeting. Kerlin's 1892 speech, which is often cited in the literature, is quoted in detail in a paper by Whitney and Schock, M.M.: "Some results of selective sterilization," 55 Proc. Annual Session Am. Assoc. for the Study of the Feeble-Minded 330 (1931) p. 331, where Kerlin says: "Whose state shall be the first to legalize oophorectomy and orchotomia for the relief and cure of radical depravity?" Whatever the actual words used, the bait was clear and Kerlin's own state was the first to pass such a law, though it failed to become law because of Governor Pennypacker's veto.

3. As quoted in Laughlin, Harry H.: Eugenical Sterilization in the United States, (1922) p. 36. See, Vetoes by the governor of bills passed by the legislature, Session of 1905, p. 26. Barr, in his book, Mental Defectives, Their History, Treatment and Training (1904), explained the purpose of such legislation: "Let asexualization be once legalized, not as a penalty for crime but a remedial measure preventing crime and tending to the future comfort and happiness of the defective; let the practice once become common for young children immediately upon being adjudged defective by competent authorities properly appointed, and the public mind will accept it as a matter of course in dealing with defectives; and as an effective means of race preservation it will come to be regarded just as is quarantine – simply a protection against ill. The fact that the practice has been perverted to selfish and vicious uses, does not affect the argument that the experience of many peoples in many ages has compelled a resort to it to preserve order in special social conditions of the times . . . " (p. 191). For a real tirade against "degenerates," see Barr's article, "Some notes on asexualization; with a report on eighteen cases," supra note 2, p. 234.
4. Goodell, William: Clinical notes on the extirpation of the ovaries for insanity, 38 Amer. J. Insanity. 249 (1882) p. 295.
5. Hofstader, Richard: Social Darwinism in American Thought (1955), Boston, Beacon Press, p. 161.
6. Ewell, Jesse: A plea for castration to prevent criminal assault, 11 Virginia Medical Semi-Monthly (11 January, 1907) p. 464.
7. Report of the Eugenics Section of the American Breeders' Association (1914), in Laughlin, Harry H.: Scope of the Committee's Work, Eugenics Record Office Bulletin No. 10A (Cold Spring Harbor, Long Island), p. 16 and passim.
8. Landman, Jacob H.: Human Sterilization; The History of the Sexual Sterilization Movement, New York, Macmillan, (1932) p. 255.
9. Just prior to Governor Sproul's veto, Dr. H. W. Mitchell, then Secretary-Treasurer of the American Medico-Psychological Association

and Superintendent of the State Hospital at Warren, wrote to Harry H. Laughlin saying that: "At this hospital we are quietly using the method occasionally, with the consent of all concerned, though the surgical treatment is given in some general hospital." Letter of October 16, 1920, as quoted in Laughlin, Harry H.: Eugenical Sterilization in the United States (1922) p. 40.

10. "Some results of selective sterilization," supra note 2 p. 332.
11. Watkins, Harvey M.: Selective Sterilization, 54 Proc. Annual Session Amer. Assoc. for the Study of the Feeble-Minded (1930), at 63. Dr. Mary M. Wolfe, Superintendent of the Laurelton State Village for Girls at Laurelton, Pa., in commenting on Dr. Watkins' paper, said that with a village composed of from 80 to 90% of moron class, "we soon came to the conclusion that selective sterilization was a measure that should be taken. Today we are sending out on parole approximately eight to ten percent of our girls. I believe we could send out thirty-five to forty percent if we had selective sterilization." *Ibid.*, p. 66. Apparently, Dr. Wolfe was not willing to act without benefit of law, unlike some other Pennsylvania administrators, notably those at Elwyn.
12. This was in answer to my inquiry about whether any further sterilizations had been performed subsequent to those reported in the 1931 Whitney-Shick article. Dr. Clark answered: "I do not know of any later statistics than those mentioned." (Letter to author, July 30, 1962.) This might tend to confirm the statistics reported in the article. Just when sterilization ended at Elwyn is not indicated. However, Dr. Clark in a letter to Dr. Jens A. Dalgaard, Director of the Bureau of Mental Health Services for Children in the State Department of Public Welfare, said that he (Clark) had advised me that there was no further information that he knew of more current than the 1931 Whitney and Shick article, and also, that he had advised me that "sterilizations were discontinued at Elwyn many years ago." (Letter dated August 27, 1962, copy sent to author by the Commissioner of Mental Health, Dr. John E. Davis, October 4, 1962.)
13. Letter to author, August 1, 1962.
14. Letter to author, August 17, 1962.
15. Letter to Dr. Dalgaard, dated September 21, 1962, copy sent to author by Commissioner Davis on October 4, 1962. The Governor was Pennypacker, supra note 3.
16. Senate Bill 322 (Providing for Emasculation of Sex Offenders), introduced in the 1961 Legislative Session, read as follows:

    Any court of common pleas of the county in which the defendant resides, is lawfully confined, or was convicted of any of the hereinafter enumerated crimes may order the emasculation of any person upon proof of either of the following: (1) That such person has been convicted of the crime of commonlaw rape or of assault with intent to ravish, or of murder committed in the perpetration of rape, or of an attempt to commit any of such crimes upon any female child under the age of sixteen years. (2) That such person has been convicted of a crime in the commission of which he engaged in sexual activity of any nature with a child under sixteen years of age and has been declared to be a constitutional psychopathic inferior by three psychiatrists and two qualified physicians acting as a board under appointment by the court hearing the case . . .

In their press release of June 19, 1961, the Philadelphia Branch of the American Civil Liberties Union denounced the measure as "vague, sweeping, carelessly drafted, and unconstitutional."

17. A twenty-four-year-old woman, Francine Rutledge, was found guilty of second degree murder for having thrown her four-day-old baby to its death down an apartment building incinerator chute the previous September. This child was her fourth illegitimate baby, the other three illegitimate children having since been placed in foster homes. According to newspaper accounts, she has an IQ of 62 and a mental age of 12, and was under great strain following the birth of her fourth child. Before sentencing her to an indefinite term in the State Correctional Institution for Women at Muncy, Judge Raymond Pace Alexander of the Court of Common Pleas (No. 7) of Philadelphia advised her to undergo a sterilization operation. He pointed out that he could not order her to be sterilized and told her, "This is a recommendation, not a compulsion . . . you ought to do it." (Philadelphia Inquirer, June 1, 1966, p. 11.) He emphasized the fact that he was not ordering sterilization, but that he was merely suggesting it "as the best solution in a situation of this sort." Following the imposition of sentence, he told her that if she underwent sterilization, he would consider releasing her in 3½ years. (New York Times, June 2, 1966, p. 37.) For a discussion of the Hernandez case in California involving a judge's offer of probation for sterilization, see this writer's article: The return of punitive sterilization proposals; current attacks on illegitimacy and the AFDC program, 3 Law and Society Review 77 (1968), fn. 3 p. 79-80.
18. Paraphrasing of story furnished by Samuel R. Rosenbaum, Esquire, of the Philadelphia Bar, and printed in Van Roden, Edward Leroy: The legal trend of sterilization in the United States, 22 Pa. Bar Assoc. Quar. 282 (1951) p. 291-92.
19. The symposium on "Eugenic Sterilization" at the 1970 annual meeting of the Pennsylvania Medical Society, which was stimulated by the request of one Pennsylvania gynecologist for enlightenment on the question of sterilizing mentally retarded girls even with parental consent, could lead to several possible avenues for an answer to this question, especially in light of the fact that Pennsylvania still does not have a sterilization law, either of a compulsory or voluntary kind. One technique, used in a recent Kentucky case (A. S. Holmes and Whitley

County Medical Society v. Allene Powers, Court of Appeals of Kentucky, Dec. 13, 1968) was a request for a declaratory judgment to define the rights and liabilities of the parties involved. Another possibility would be for one of the state health or mental health officials to ask the state Attorney General for an official opinion on the subject of sterilization. An earlier discussion of the subject of voluntary sterilization in Pennsylvania can be found in Challener, William A., Jr.: The law of sexual sterilization in Pennsylvania, 57 Dickinson Law Rev. 298 (1952) and the Pennsylvania case of Shaheen v. Knight, 6 Lyc. 19, 11 Pa. D. & C. 2d 41 (1957). For a critical discussion of the use of probate court orders for sterilization in a neighboring state (Ohio), also without any specific sterilization statute, see the note on In re Simpson, 180 N. E. 2d 206 (Ohio P. Ct. 1962) in 61 Mich. L. Rev. 1359 (1963). For a discussion of the current status of state eugenic sterilization laws, see the author's article: The psychiatrist as public administrator; Case in point: State sterilization laws, 38 Amer. J. Orthopsychiatry 76 (1968).

20. A pointed example of a legislative proposal that failed in the Hawaii Legislative in 1970 is the following bill entitled "A Bill for an Act Relating to Population Control" (S. B. No. 1421-70), which read:

    SECTION 1. The legislature finds: (1) that population growth is the most serious and most challenging problem for mankind today; (2) that the time necessary for the population of the world to double is now about thirty-five years; (3) that the "death rate solution" by war, famine, or pestilence is an unacceptable destructive solution to the problem of population growth; and (4) that population control is an acceptable humanitarian solution to the problem of population growth. The purpose of this Act is to control the population size of this State by a program of birth regulation.

    SECTION 2. Every physician attending a woman resident of this State at the time she is giving birth in the State shall, if the woman has two or more living children, perform such medical technique or operation as will render the woman sterile.

21. For a discussion of these punitive proposals, see the author's article: The return of punitive sterilization proposals; current attacks on illegitimacy and the AFDC program, 3 Law and Society Review 77 (1968).
22. Buck v. Bell, 274 U.S. 200 (1927), at 207. See chapter by Giannella, D., supra.
23. One prestigious organization that raised serious questions about the use of compulsory sterilization laws was the American Neurological Association in its landmark report, Eugenical Sterilization; A Reorientation of the Problem (1936). The Task Force on Law of the President's Panel on Mental Retardation (1963) came to the following conclusion: "There are serious questions about both the validity of the scientific assumptions on which these laws were based and the way in which it is decided who

should be sterilized . . . We do not take a position on whether sterilization can ever be ethically justified. Our recommendations are limited to urging that the operation not be allowed to result from misjudgment as to its scientific need or from inadequate opportunity for administrative and judicial review." (Report of the Task Force on Law, pp. 22-23). It should be noted that the return of punitive sterilization proposals discussed at the end of the paper are really not eugenic in character in that genetic factors are not at issue. Much of American eugenics history is marked by this mixing of eugenic with noneugenic expectations. The current call by some persons for sterilization as a means for stopping the so-called population explosion is probably closer in its intent to the old "cacogenic" (or socially inadequate) categories than it is to anything remotely genetic. One of the things that surprised this writer about the urging of some groups in Pennsylvania for a reassessment of eugenic sterilization was the belief that eugenic sterilization in America was pretty much of a dead issue, particularly in the light of the data that I collected on Pennsylvania sterilization history. Pennypacker, quite obviously, has not had the last say on this matter.

Chapter IV

EUGENIC STERILIZATION: MEDICAL-MORAL CONSIDERATIONS

FRANK J. AYD, Jr.

IN the past two decades we have witnessed a growing clamor for legalization of abortion and for more widespread use of voluntary sterilization. Perhaps, therefore, a panoramic view of our contemporary medical-moral culture should be presented first, for its is in this setting that the present push for sterilization exists.

Twentieth century man has become progressively less theocentric and more materialistic and egocentric. To some, God is dead; to others, He is irrelevant. A new religion – scientism – has sprung up, the high priests of which are scientists and humanists. They have erected a new hierarchy of values. The *quality* of life, they say, is more important than the sanctity of life. They tell us that there is no necessity to be preoccupied with or dedicated to man's spiritual welfare since there is no scientific evidence or proof that there is a soul or a supernatural destiny. Only the welfare of mankind is important. This takes precedence over the welfare of the individual, and when necessary, individuals can and must be sacrificed.

SITUATION ETHICS

The high priests of scientism have rejected traditional Judeo-Christian morality as obsolete and unworthy of adherence to by man enlightened and freed by the discoveries and attainments of science. Ethics, they say, is situational. There is a new morality which holds that every human act – even murder – is good, if it is motivated by *love*. Premarital and extramarital sexual relations, they contend, are healthy and morally permissible. So, too, are homosexual relations between consenting adults. Pornography does not pollute the mind or generate unhealthy carnal desires. Therefore, they aver, pornography should not be restricted but easily accessible to all – young and old. Man is free to indulge in sexual pleasures in any way he wishes. There is no reason to feel guilty or to be inhibited. Sex is to be enjoyed.

Just as there are new sexual mores, so too are there new reproductive mores, because science has enabled man to be infertile if he wills, and to procreate in new ways. Sexual relations now are primarily for the giving of love and not for the transmission of life.

Reproduction can and should be controlled artificially or avoided completely. Man may be voluntarily sterilized by drugs or by surgery. He may change his sex by drugs and surgery. He need not marry to have children. If he marries, divorce should not be regulated by law but be a free choice of either partner. If he desires children, he should want no more than one or two, as his contribution to the avoidance of overpopulation. Furthermore, only superior children should be desired and conceived, not primarily because of innate paternal or maternal urgings but because the couple desires to perform a service to mankind by having as well-endowed children as possible.

Man can control the biological makeup of his near and remote descendents and because he can, he should be proud to make an exemplary contribution to humanity. Men and women with superior qualifications should contribute sperm and ova to banks. The sperm from these banks would be used for artificial insemination or to fertilize outside the body ova withdrawn from the same banks. Women, if they wish, may conceive but will not have to bear and deliver their children. They may be impregnated by artificial insemination, or have their ova fertilized in a test tube, and, in either case, hire another woman, who has been synthetically prepared to be a substitute mother, to bear and deliver a baby transplanted to her womb. Since human sexuality is not confined to copulation for reproduction, men can be fathers and women mothers without ever having conjugal relations. Hence, there is no need for marriage. Single men and women can have children. Technological progress is making the traditional concept of the family obsolete.

A PROMISED UTOPIA

Scientism promises utopia, a synthetic heaven on earth made possible by chemical, biological, and technological developments. There is no need for God or outmoded religions. There is no need for self-restraint, unless this is dictated by the *superior* objectives of scientism, such as the attainment of intellectually and physically supermen, immune to disease or illnesses, capable of unlimited longevity and subject to only accidental death or deliberate self-extinction.

There is an intoxication with the powers generated by science and technology, and with the promises of the future. No wonder so many men arrogantly declare that there is no limit to what man can do. No wonder they are impatient with moral restraints. They want everyone to march to the beat of the scientific and technological drummer. Accept their godless philosophy, give them free rein, do as they bid and they assure us that mankind will be blessed with unparalleled material accomplishments. Yet, there was more than acid sarcasm in his words when Voltaire said:

> "I would have princes and their ministers acknowledge a God – nay, more, a God who punishes and pardons. Without this restraint I should consider them ferocious animals who, to be sure, would not eat me just after a full meal, but certainly would devour me were I to fall into their clutches when they are hungry, and who, after they had picked my bones, would not have the least idea that they had done anything wrong."

Suppose the glorious promises of scientists should become realities! It is gratuitous to assume that they would make man more human, more virtuous, happier, and more dedicated to achieving his supernatural destiny. Without moral progress, as Edward Shils asks:

> "How is the human race as we have known it, with all its deficiencies, to be protected from the murderous and manipulative wickedness of some of its members and from the passionate curiosity and the scientific and technological genius of others? How far does morality allow us to intervene into reproduction, the course of life, and the

> constitution of individuality and privacy? Under what conditions is the modification of temperament and individuality by *contrived intervention* permissible?"

And I would ask: What kind of a society is it that prefers to obliterate suffering rather than relieve it?

POPULATION GROWTH

Much publicity has been given to the world's population growth. Experts warn that if the growth rate is not soon reduced to zero, the human race will destroy itself by an epidemic of unrestrained breeding. Experts also agree that the oral contraceptives, the intrauterine devices and all other non-surgical methods of family planning are not the solution to the population crisis. The main obstacles to curbing population increase are not technological but human. People, the experts insist, must be convinced that they should not have more than two children. At present, they are not. They are using contraception to space children until they have the number they desire, which for Americans is three or four. This being so, those who ardently want a rapid reduction in population are clamoring for two things. First, no laws restricting abortion or a federal law guaranteeing the right of any woman to have an abortion whenever she wishes. Second, state or federal legislation requiring sex education in schools which includes discussion of the need for regulating the birth rate and the techniques of birth control. Liberal abortion laws are needed now, we are told, because only abortion will quickly lower the birth rate and we cannot wait for education to motivate people to limit family size through contraception. However, as everyone knows, but may not realize why, there are very active campaigns now going on to make people *abortion-minded* and ultimately to create a universal *contraceptive mentality* in our youth.

Whether pro-abortionists will achieve their objectives remains to be seen. In the meantime, I contend that advocates of voluntary eugenic sterilization are doing exactly what the pro-abortionists have done. At this time, they are not publicly mentioning their true objective which is the widespread use of voluntary sterilization for contraceptive purposes, not as a part of family planning but as an integral part of population planning for Americans.

Voluntary sterilization advocates know that to have their concepts accepted in our democratic society they must pass through a series of stages. According to Rufus E. Miles, Jr., Chairman of the Board of Trustees of The Population Reference Bureau, these are:

1. A few individuals *crying out in the wilderness.*
2. Small groups, working together informally at first and then in small struggling organizations, sometimes with the help of venture capital from foundations.
3. Mass media coverage, slight at first but slowly increasing thereby achieving respectability for the subject as a topic for general discussion; more foundation support; the beginnings of university interest.
4. Bills introduced into Congress by a few legislators, designed to open the attack on the identified problem.
5. Crystallization of opposition to the specific proposals, resulting in unavoidable delay in achieving further progress.
6. Appointment of one or more high-level advisory commissions of distinguished citizens to make policy recommendations and win public support for legislation.
7. Serious Congressional hearings, culminating in legislation usually of modest scope and funding.
8. Increasing acceptance of the legislation, its strengthening in structural and financial terms, and its incorporation into the institutions and mores of society.

Voluntary sterilization supporters are at stage two, moving toward stage three.

In the early days of the movement to have abortion on demand, pro-abortionists had one thing in their favor; namely that therapeutic abortion was legal in many states. Their task was to enlarge the concept of what is therapeutic and by emotional propaganda to persuade legislators and the public that abortion is acceptable because it is not the destruction of a human life, but of a *potential* person. The fetus, they say, is not a person, but rather tissue with the potentiality, in most cases, for becoming a person.

By unrelenting campaigns of persuasion pro-abortionists have succeeded in having the concept of what is therapeutic broadened. It now includes physical, psychiatric, and eugenic factors, and to some, social and economic factors as well.

Supporters of voluntary sterilization for contraceptive purposes are doing exactly what pro-abortionists have done. They are capitalizing on the fact that over half of our states have sterilization laws that provide for the compulsory and/or voluntary sterilization of mental defectives and others. They know that the existence of such statutes can affect public policy, so that it would favor voluntary sterilization, at least on eugenic grounds. Furthermore, some state laws contain a provision stating that nothing in the statute shall prevent a sterilization for therapeutic reasons.

With these legislative precedents favorable to their cause and aware of the contemporary changing mores, proponents of voluntary sterilization for contraceptive purposes and for population planning have embarked on a campaign to broaden the concept of what constitutes therapeutic sterilization. They also are employing the same emotional appeals that the pro-abortionists used so effectively to win public and legislators' support for legalization of voluntary sterilization. Some ardent advocates of sterilization for population control hope to ultimately have compulsory sterilization to help us achieve a zero population growth.

Significantly, today's sterilization supporters are relatively silent on eugenic sterilization as a means of racial improvement. They know that there is little or no scientific foundation for this. Instead, just as proabortionists have advocated abortion to prevent the birth of physically deformed or mentally abnormal children, sterilizationists are urging this operation to prevent the conception of such children. They are saying that sterilization is a simple and safe operation. They are telling the public that it is reversible in many cases. They are claiming that it is the ideal method of contraception because it requires only a single act to render a person infertile for a prolonged period. They also are claiming that as a result of the advances in genetic diagnosis and the increasing use of genetic counselling, more and more people will want to be sterilized and that this should be their legal right. For such people, of course, voluntary eugenic sterilization is in fact contraceptive sterilization.

To many of those concerned with the *quality* of life, mental retardates are *inferior* members of society. At present they would oppose their deliberate extermination as too drastic a solution to the myriad problems they pose. However, their fervor for population *quality* and their concern about fitness for parenthood compels them to advocate eugenic and punitive sterilization of the mentally retarded. Such enlightened prevention of the reproduction of the mentally defective, they claim, would reduce the incidence of retardation and lower public welfare costs.

Those who want sterilization – voluntary or compulsory – for the mentally defective argue that society must look upon germ-plasm as belonging to society and not merely to the individual who carries it. This gives society the right to ignore individual civil liberties and human rights. Hence, they urge a public policy on sterilization as a birth control measure and, in the past decade, there has been a gradual increase in the clamor for punitive sterilization of the mentally retarded. Some public health officials have proposed that courts should order sterilization of mentally defective mothers to halt the production line of unwanted children they bear. Some judges have ordered sterilization of young mentally retarded women so they could remain in the community without burdening society with the children they may have if they are not rendered sterile.

It is evident that voluntary eugenic sterilization is being urged not just for the presumed benefit of individuals but as a means of solving some of our social problems. It also is evident that some sterilization advocates are not primarily interested in the welfare of individuals whom they wish to be sterilized. They really desire to use sterilization as a birth control measure to reduce the population growth and thus allay their anxieties about what they call the population bomb. They also want sterilization to raise the *quality* of life and thus to have the satisfaction of feeling they have contributed to the betterment of mankind.

These are laudable goals but should they be achieved, if they can, through voluntary and compulsory sterilization? Sterilization is no more a medical problem just because the physician wields the scalpel than chemical warfare is purely a problem for pilots because they press the lever releasing the chemical. We must consider the morality or immorality of sterilization.

A Hospital Compendium: A Guide to Jewish Moral and Religious Principles in Hospital Practice contains two statements on sterilization. These are:

> 1. Man has been commanded in the Bible to *be fruitful and multiply and fill the earth.* This duty marks the first of the 613 commandments which serve as a Divine law enjoined on the Jewish people. A Jew is not discharged of this obligation until he has begotten at least one son and one daughter. By the act he is viewed as *a partner with God in the Creation.*
>
> The fundamental premises upon which are based detailed Jewish teachings on the moral problems connected with the inception of life serve to mold the Jewish view on sterilization, contraception, and abortion. Any deliberate interference with the natural process of generation is viewed as an offense, unless necessitated by medical indications. The sterilization of the male is viewed as a particularly serious offense and is never permitted unless it is necessitated as part of surgical procedure designed as a life-saving medical intervention. Such occasion arises, for example, in prostatectomies. Although sterilization of the woman is viewed somewhat more leniently by Jewish law, no surgical procedure which may interfere with the procreative ability of the woman may be undertaken without careful consultation with a competent rabbinic authority.
>
> 2. Surgical or physical impairment of the reproductive organs of any living creature violates Jewish law, except in cases of urgent medical necessity. In the case of males, only a risk of life can justify such procedures; hence, unless medically demanded, the ligation of the vas deferens during prostatectomy should be avoided. The prohibition against impairing the male reproductive organs and functions is unrelated to man's fertility. It applies even to a man known to be sterile or impotent, whether by reason of age or physiological aberration.
>
> The sterilization of females is permitted when medically imperative. Rabbinic opinion, as well as the consent of the husband, should always be secured prior to considering ligation of the fallopian tubes or other surgical procedure leading to sterility. (Commission on Synagogue Relations, Federation of Jewish Philanthropies of New York, Inc., 130 East 59th Street, New York, New York 10022).

Traditionally, Christian theology has condemned all forms of direct sterilization, whether compulsory or voluntary. This condemnation is based on the Christian view of man as compounded of matter and spirit and destined to live in society. He has inherent and inviolate liberties and restrictions that are imposed on him by his very nature as man and as a social being. From his Creator he has received life, the freedom to do good or evil, and the obligation of preserving his life and bodily integrity. Because man is the custodian and user, but not the owner, of his life and body, the uses he can make of these are limited and governed by the natural law. He cannot confer powers on others that he does not possess. Hence, there are things to which he cannot consent without violating the natural law. On the other hand, under certain circumstances from a moral standpoint, he can be sterilized licitly, for example, therapeutic sterilization is morally justified if it is the *only* means of securing the welfare of the body as a whole. It follows, however, that he cannot do this without knowing and consenting to it.

In view of these facts, it is not, nor can it ever be, permissible to sterilize a person without his being aware of and permitting it. The consent to sterilization must be an informed consent. Obtaining an informed consent from many mental defectives and mentally ill individuals is very difficult, if not impossible. They are incapable of giving a true consent, since they would be unable to grasp all the implications of a sterilizing operation. Since a sterilization program for these people must indeed be voluntary, there must be safeguards against the exercise of undue influence.

To medicine's shame, some physicians are sterilizing women without their knowledge and consent or after coercing them to consent. Most of these victims of the tyranny of silent or coerced sterilization are from the socio-economic underprivileged classes of American society. Few of them are mentally retarded. Most of them are *welfare* cases. This is becoming a matter of grave concern to the black community. They consider it genocide and their objections to it are becoming more vociferous.

The Roman Catholic church does not countenance voluntary sterilization for eugenic reasons. A Church of England Committee has stated that a normal couple who knows that one partner is a carrier of defect may regard the sexual organ as *diseased,* in that the genes it secretes are defective, and given circumstances of *necessity,* may resort to sterilization. Norman St. John-Stevas rebuts this argument by pointing out that the organ is not diseased in relation to the parental body but only to the hypothetical child. What constitutes *necessity?* Here the couple, he says, have two alternatives; they can abstain or they can use contraceptives. Accordingly, sterilization would not be justified in their case.

There are those who advocate sterilization of subnormal couples who are incapable of using effectively currently available contraceptives and who the advocates believe would be inadequate parents whether or not a child they may have is normal. They contend that no child should be born without the birthright of a sound mind in a sound body and under proper conditions. Since subnormal couples may have a defective child or be inadequate parents, they say the child has the right not to be conceived. Paradoxically, these same advocates of sterilization argue that a conceived child has no rights and hence can be aborted, but that a non-existent child does have the right to have the initiation of his life prevented by the sterilization of his potential parents. They would have a more rational argument if they contended, as all religious denominations do, that no person has a moral right to procreate if he has no reasonable assurance that he can beget healthy offspring and make reasonable provision for them.

One obligation of citizenship is the duty to engage in law-making. The present clamor for sterilization for contraceptive purposes and as an integral part of a possible national population policy compels us to decide if this shall become a fundamental ingredient in our civilization. One thing we must be sure of is that in any statute on sterilization there should be provision for *conscientious objection* on the part of physicians, nurses, and other hospital staff members regardless of their religious persuasion.

One objective of this paper is to focus attention on certain contemporary anti-life trends. We may have abortion on demand. We are being asked to take a permissive attitude towards sterilization. There are those who now countenance infanticide, which incidentally we already have under some of our liberal abortion laws which permit abortion after the medically acknowledged time of viability. (In Maryland, for example, abortion is legal up to twenty-six weeks gestation). We are witnessing a growing campaign for euthanasia.

CONCLUSION

It is my sincere hope that the views I have expressed will not be misinterpreted as a condemnation of the advocates of sterilization or as a lack of sympathy for and understanding of their earnest and genuine motives to improve man and his condition. Our basic aims are identical. It is primarily our divergent concepts of human values and the means advocated to achieve our goals which provoke disagreement. Christians submit that what is envisioned must be sought in ways consonant with man's total nature. We are as desirous as anyone else to enable contemporary man to achieve his natural fulfillment to the fullest measure possible; to improve the physical quality of life; to relieve pain and unhappiness; and to enhance human powers. Yet, we cannot condone the views of those who seek these objectives from aspirations to omnipotence, and who wish to manipulate the individual and to intrude on the privacy of human beings with utter disregard of their dignity as persons, their inalienable rights. We hold that human life is sacred. If life is not sacred, nothing else could be. The fact that many human beings act contrarily, or do not apprehend it, does not impugn the sacredness of life. Many who could do extraordinary harm to humanity do not exploit their present powers, because they revere life. Man may know more, but he never has been in greater peril of being dehumanized.

Chapter V

GENERAL REMARKS ON MEDICAL ASPECTS OF MALE AND FEMALE STERILIZATION

ALAN F. GUTTMACHER

MALE sterilization through castration is very ancient; there are no historical records that reveal the first instance because castration was commonly performed by the victorious warring tribe on the vanquished in order to make slaves of the captured men and make them amenable slaves. Castration was also carried out until late in the nineteenth century to create voices for the Sistine choir. Male sterilization was practiced by primitive peoples by creating an artificial hypospadias; thus the male ejaculate dribbled forth from the base of the penis and not from the tip and, therefore, he could not impregnate. Sterilization by vasectomy is very recent. The literature informs us that in Indiana in the last part of the nineteenth century a misguided physician named Sharp who was in charge of a home for feeble-minded boys first carried out vasectomy in an attempt to cure them from masturbating. This, as can be anticipated, had very little effect on their behavior.

The history of female sterilization is beclouded by legends. It is claimed that the maidens, who bore the image of Atilla in the religious processions in classic Greece, were previously oophorectomized. There is little to support this. On the face of it, it seems virtually impossible for an abdominal procedure at this antique period would have been too hazardous to be feasible. The first verified oophorectomy was carried out by Edinburgh-trained Ephraim McDowell in 1809. This was a most extraordinary procedure; McDowell was called in consultation some sixty miles from his office to see a woman deep in the back woods of

Kentucky who had a vast abdominal tumor. He assumed it to be an ovarian cyst and told her that there were two possibilities, one was death and the other to ride sixty miles on horseback to his office and submit to an operation. The very brave woman chose the latter. Dr. McDowell removed the vast tumor, the first time in the history of medicine that we can be certain an ovary was removed. Five days later, with the sutured pedicle of the ovary protruding from the lower part of the wound, the woman returned sixty miles on horseback.

Bilateral oophorectomy dates from a Philadelphia surgeon, Atlee, in 1843. It later became a very popular procedure. The ovary was thought to have very little function except to create children. Its endocrine necessity was badly understood and it is said that at one time in American medicine a large slop jar was kept in the operating room into which excised ovaries were deposited. Such a distinguished physician as Meigs, Professor of Obstetrics and Gynecology in Philadelphia, thought that orgasm was the prime cause of female cancer. He reasoned that if orgasm could be eliminated by oophorectomy the chance for female cancer would thus be reduced. There was also the syndrome of *ovariomania* which ordinarily expressed itself as nymphomania or persistent masturbation, both of which dictated bilateral oophorectomy. Many pairs of ovaries found their way into the bucket for these and other reasons.

The first tubal sterilization was performed by Lungren of Toledo, Ohio in 1880. The operation has survived these ninety years and its use has been greatly increased since then. There are many types of tubectomy or salpingectomy.

Let us now examine the place of tubectomy and vasectomy in modern medicine. I was trained at The Johns Hopkins in Baltimore and had the great advantage of being taught by a very remarkable man, a trail blazer, a man of great character and intestinal fortitude, Dr. J. Whittridge Williams, the Chief, as we referred to him. Dr. Williams published a paper on the sterilization of eighteen or twenty feeble-minded women in the Journal of the American Medical Association about 1925. Thus I entered medicine with the understanding that sterilization of the female had a place. Its common indication was serious mental retardation.

I well remember one such patient on whom I did a Cesarean section and tubal ligation. I was resident at the time. She was operated upon about eight in the morning and as customary I made midnight rounds with a flashlight going from bed to bed. I recall my astonishment when I reached her bedside about fourteen hours after surgery. I though I saw an apparition because here sat this huge woman on a chair twiddling her thumbs over her bandaged abdomen. This was years before the current practice of early ambulation. I became terrified and thought we would find the wound gaping wide open. Several strong bodies hoisted her back into bed and much to my relief inspection of her incision showed no damage had been done. As a physician in private practice I have done occasional sterilizations on adolescent females brought to me by their parents for sterilization because of serious mental retardation. When I did this I had no fear of legality, if I first obtained signed permission of the parents and psychiatric evidence on the chart of the young woman's inability ever to take proper care of a family. However, one had to decide whether to sterilize by hysterectomy or by tubal ligation, because in some cases the patient is so vastly subnormal that both the consulting psychiatrist and family feel that she is incapable of proper menstrual hygiene. If this is the decision, one leaves the ovaries in and removes the uterus. I feel that oophorectomy is never wise and proper for adolescent sterilization. One carries out sterilization to relieve the individual of the burden of caring for children – not because one fears transmission of the defect to offspring. The few adolescents I have sterilized have been brought to me by normal, concerned parents. In my limited experience the father is usually a physician who fully understands the situation. May I once more stress the fact that such sterilizations are not done on eugenic grounds. It is done because of the inability of the young women to ever be able to cope with a family and to give children a proper environment. I am *lukewarm* about eugenic sterilizations on the basis of the transmission of defects, particularly mental defects. The exception is cases in which chromosomal abnormalities can be demonstrated by karyograms, which is entirely different.

Sterilization is widely applied today to the female for many reasons other than eugenic or genetic. I am not going to present a catalogue or listing of medical conditions which render pregnancy hazardous involving the heart, lung or various other organs because these somatic conditions are well understood. Studies done independently in the late 30's and early 40's by Jacob Yerushalmy and Nicholson Eastman demonstrated clearly that both fetal and maternal mortality rise quite sharply with the eighth and later children. American medicine, therefore, suddenly accepted the concept that great multiparity was in itself an indication for sterilization, because sterilization of the grande multipara would reduce both maternal and neonatal mortality as well as diminish mental retardation and other abnormalities associated with children born to greatly parous women, particularly those in the older age group.

At Johns Hopkins, where I served for many years, we put into effect the Para Eight rule: any woman could be sterilized who had had eight births. When I went to Mt. Sinai in New York in 1952 and took over the obstetric-gynecology service, I found that sterilization for the private patient was relatively easy to acquire, but sterilization for the ward patient was unobtainable. In order to equalize sterilization on the two services I put into effect a rule which I am now not very proud of, but in 1952 believe it or not, it was a radical rule. Any woman with her sixth living child irrespective of age could be sterilized, also any woman age thirty to thirty-five with five living children and any woman thirty-five or more with her fourth living child. This became known as the law from Mt. Sinai. For far too long it was rather slavishly followed in American medicine. I regret to report that several hospitals still follow the law from Mt. Sinai. To be sure it has been modified in many institutions.

Then there is the *numbers game.* Some hospitals have chosen 120, some 135. One multiplies the mother's age by her living children; a woman of 30 with a fourth living child qualifies for sterilization at hospital A, but if she is 29 years old with a fourth living child, that is only 116, and therefore she does not qualify. Both the Mt. Sinai formula and the multiplication of age is being abandoned in most areas, I am happy to say. In American medicine a stronger relationship is developing between a doctor and a patient. If the physician feels that sterilization serves the purposes of the couple, even though they only have two or three children, in most institutions sterilization will now be carried out post-delivery.

An interesting situation recently occurred at a hospital in Northern Westchester, New York. A woman who was quite young delivered her third child. Her physician was willing to honor her request for sterilization; but the hospital would not permit it. She brought suit and before the suit could be tried the hospital capitulated, sterilized her, and eliminated rules such as the *numbers game.* A far more liberal attitude toward female sterilization is gradually sweeping the country.

There is some nationwide data on the incidence of sterilization gathered from a sample survey of six thousand white married wives, age eighteen to thirty-nine. It appears that 8% of the marriages have been rendered infertile through surgery for that purpose. In 3% the husband had a vasectomy and in 5% the wife a tubectomy. Thus; one in twenty women in the United States has been sterilized to terminate fertility. Many female sterilizations are done at the time of Cesarean section. Established obstetrical routine calls for sterilization at the third Cesarean section unless the hospital is Catholic or the patient objects. Frequently sterilization is offered at the second Cesarean section. There have been several studies concerning sequelae to female sterilization. Barnes and Zuspan did a study involving about 250 female sterilizations. They found complete acceptance in the private patient segment of 98%; the rate of acceptance by the ward population was slightly lower. I did a study a few years earlier on 157 women who had been sterilized on the ward service of the Mt. Sinai Hospital in New York. They were all post-partum sterilizations unassociated with Cesarean section. One to five years later the question was asked, "Would you have had the operation if you had known the results?" In other words would you do it over again if you had the choice. Seven of the 157 answered "no". Their reasons to the surveyor seemed trivial and unimportant, but to the patient probably quite important. One woman had six daughters and she was sterilized after the birth of her first son. She said if she had known how much pleasure she would get out of a son she would not have been sterilized. Another woman was living in very cramped quarters and shortly after the birth of the child, at which time she was sterilized, she was given larger quarters by the housing authority. She said if she had known she would get a larger flat she would not have been sterilized. These are very real issues to the people involved but as physicians it is very difficult for us to anticipate such problems. I came to the conclusion that female sterilization is an extraordinarily successful procedure almost universally applauded by the people who elect it. Occasionally the lack of reversibility causes a problem. Four young Puerto Rican women who had been sterilized in Puerto Rico where 30% of the female population by the age of forty-five have been sterilized came to Mt. Sinai to have the operation undone. In each instance they had remarried. Two of them had been sterilized prior to their first marriage, since the men they were going to marry insisted they be sterilized before marriage was consummated. Later they were divorced. Our operation was successful in two of the four, a 50% average, which is about the best you can do. Any higher claim is not dealing with reality.

Female sterilization as a method of population control is not widely used around the world because hospital beds are so difficult to obtain. For example, in 1967 India reported one and a half million sterilizations, 1.4 million by vasectomy and 100,000 by tubectomy. For population control the main sterilization technique has been vasectomy. However, female sterilization is now being done by laparoscopy, which may make a difference. The laparoscope is a kind of periscope which is inserted through the abdominal wall and by a series of lights and lenses one gets excellent visualization of the intra-abdominal and pelvic organs. By inserting a second peritoneoscope, an operating peritoneoscope, an actual cautery can be introduced and the fallopian tubes severed by cauterization. This technique is being used in many places around the world. Dr. Daphne Chun of Hong Kong has reported a large series. Dr. Robert Neuwirth at Bronx-Lebanon Hospital in New York City has also done a large number. Dr. Martin Clyman of New York is carrying out sterilization through the vaginal culdoscope.

At The Johns Hopkins in Baltimore female sterilization via the peritoneoscope is being done on an ambulatory basis. The patient is admitted to the hospital without breakfast in the morning and given a general anesthetic. Peritoneoscopic sterilization is carried out and she is discharged the same afternoon. Thus the logistics of female sterilization is being radically simplified. This may have application to population control in distant countries. AID is hosting two month fellowships to foreign physicians working in population control so that they may learn peritoneoscopic and culdoscopic techniques of tubectomy and apply them to their home areas. During my medical career I have seen the *stone wall* break and fragment in respect to female sterilization. Earlier in my medical career it was almost unattainable, that is if you were a normal individual who simply desired sterilization for termination of fertility. Now there is little resistance. Undoubtedly there are still communities in which resistance remains and similarly hospitals in which there is resistance, but it is gradually being eroded.

Vasectomy is also becoming much more widely performed. Senator Gaylord Nelson's birth control pill hearings in Washington unintentionally became a great boon to sterilization. Men and women, particularly intelligent men and women, became alarmed about the birth control pill. This reaction had a great impact on encouraging sterilization.

I am President of Planned Parenthood, a post of which I am proud and I am especially proud that nine of the Planned Parenthood clinics are performing vasectomies within their own walls. In Houston they are doing twenty cases twice a week. Intramural clinics of this type permit a relatively low fee. The Margaret Sanger Research Bureau in New York established the first Planned Parenthood intramural vasectomy clinic. It now meets twice a week and the two operating tables permit them to do fifteen patients each session. Dr. Sobrero, Director, uses it to train his nine foreign fellows in vasectomy techniques so by the time they return to their respective countries they will have gained familiarity with the procedure. In addition to the nine, four more clinics are contemplating vasectomies on their own premises before the first of January, 1971. One can prophesy that within two or three years more than half of the Planned Parenthood clinics will be performing vasectomies. One Planned Parenthood clinic is doing abortions. In Syracuse, Dr. Penfield, is doing twelve abortions on early pregnancies a week. Syracuse is also planning to commence sterilizations. With such diversity, it will actually become a family planning clinic being able to offer any method of conception control that is indicated. No doubt tubectomy by peritoneoscopy will be added at not too distant a date.

Vasectomy is used extensively in India, where it is a very simple procedure. When I was there a little less than two years ago, I witnessed a vasectomy in one of the health centers. Architecturally the center resembled a chicken coop with a cement floor and a kitchen table. Overhead, the light was about forty watts. The nurse and doctor were scrubbing up. Finally an old man came in leaning on a cane, gnarled and limping. They hoisted the chap up on the table, swished up his gown and scrubbed him with green soap. The operation was done expertly. I was much confused by the disparity between the procedure and the man's apparent age. So I said to the doctor, "Ask the chap how old he is, will you?" So in his best Urdu or Hindi he asked the fellow how old he was. He had no idea how old he was; it was obviously a foolish question. So then I told him to ask how many children he had. The answer came back, "Nine." Here was a cocky fellow who had already sired nine children, in his 60's or 70's, who had come for vasectomy. This presents one of the problems of population control. In India by the time a man is vasectomized he has four and one-half living children. And since marriage occurs at an early age, each generation is about twenty years. If each twenty years one replaces two parents with four and one-half children, you can easily see that no ground is gained.

In regard to discovering a fully reversible sterilization procedure for both the male and female, much study and research is going on without magnificent success. There are two active male programs. When I go to India I hope to investigate the program in Bombay which applies a small tantalum metal clip to the vas deferens. The clip is closed sufficiently to obliterate the lumen of the vas but, hopefully, not sufficiently to cause tissue necrosis. Then theoretically one could remove the clip after an X number of years and fertility would be restored.

In South Korea, Dr. Lee introduces a nylon thread through the intact cord into the lumen of the vas and then brings the thread out through the skin. On either end of the thread a tiny lead shot is buried beneath the skin. If the man wants his fertility restored one cuts off one of the lead shots and pulls out the thread. This technique has not proven as successful as initially thought. It reduces but does not eliminate fertility. With decided improvement in the safety and simplicity of contraception, it is possible that we shall see vanishing use of sterilization for the married couple. If the exciting new work on the prostaglandins comes into fruition, and if by inserting into the vagina a suppository with fifty milligrams of prostaglandin F_2 Alpha whenever a woman is two to five days overdue on her menses causing her period to come on promptly, pregnant or not, we would then have a simple, safe and effective method of post-conceptive contraception. Then sterilization would become much less urgent among couples of normal intelligence. The problem would remain with the mentally subnormal.

Chapter VI

EUGENIC STERILIZATION AND THE LAW

DONALD GIANNELLA

IN considering the legality of eugenic sterilization, one might be tempted to both start and end with the famous phrase of Mr. Justice Holmes in *Buck v. Bell* that *three generations of imbeciles are enough* (1). In the *Buck* case the Supreme Court upheld the constitutionality of a Virginia law providing for the compulsory sterilization of mental defectives. This 1927 decision would seem to establish the legality of eugenic sterilization generally and leave open for examination and argument only the desirability of a social policy compelling or encouraging such operations. However, the *Buck* case has come in for some heavy criticism from legal commentators down through the years and is a decision that bears re-examination.

It should also be noted that the *Buck* case involved a legislative enactment. Approximately half the states do not have such statutes. The question then arises whether in such states a court may authorize the sterilization of a legally incompetent person. There is some division of legal opinion on this point, but the more authoritative position holds that the court has no such power in the absence of a statute expressly providing for the sterilization of mental defectives.

Although the topic, eugenic sterilization, primarily raises the problem of the mental retardate or defective who cannot give a legally valid consent to the operation, it is broad enough to include the case of the mentally competent adult who might seek sterilization because he is concerned about transmitting an inheritable disease, such as diabetes, to his descendants. Even after *Buck v. Bell* there was considerable doubt whether such a voluntary operation would be legal. Voluntary sterilization for therapeutic reasons was clearly legal, as in the case of an invalid whose health would be seriously imperiled by pregnancy; but voluntary sterilization for the purpose of birth control was generally considered unlawful (2). There was some uncertainty in predicting whether a court would liken eugenic sterilization to a therapeutic operation or regard it simply as an improper form of birth control. Developments in recent years indicate that voluntary sterilization as a form of birth control is now quite clearly legal. However, some of these developments have added to the growing doubts about the continuing vitality of *Buck v. Bell.*

In discussing eugenic sterilization, I will take up first the legality of voluntary sterilization of competent adults, then the case of a court order authorizing such an operation on a legally incompetent person in the absence of a statute, and finally the constitutionality of statutes providing for the sterilization of mental retardates and defectives who cannot personally give a legally valid consent.

VOLUNTARY STERILIZATION

Thirty years ago a scholar trained both as a lawyer and a doctor confidently maintained that non-therapeutic sterilization was illegal under most contemporary criminal statutes prohibiting maiming since it involved a "wounding for an asocial purpose . . . and the permanent destruction of a socially useful bodily function" (3). He also thought that voluntary sterilization constituted both a criminal and civil assault and battery (4). He arrived at this conclusion on the ground that the operation was contrary to the then prevailing public policy prohibiting the advertising of contraceptives and, in some states, even their sale and use. Because of this public policy, he assumed that the consent of the patient to the operation would be legally ineffective.

The last thirty years have seen two developments that would appear to establish the legality of voluntary sterilization. First, concern about world over-population and the impact of population growth on environmental pollution and depletion of natural resources in the United States has led to a national policy favoring birth control (5). Consequently, contraception is now viewed in a new and favorable light. A lower court in Pennsylvania gave early recognition to this change in policy by holding in 1956 that an agreement to perform a non-therapeutic sterilization procedure was not contrary to public policy (6). The court did recognize that many in our society still look unfavorably upon sterilization but because of the wide division of opinion it refused to find the operation contrary to public policy.

A second important development favoring voluntary sterilization was the emergence of a new constitutional right to privacy in sex and family planning matters. In 1965 the United States Supreme Court decided the *Griswold* case that struck down the Connecticut statute forbidding the use of contraceptives (7). Although there was some confusion and disagreement among the members of the Supreme Court regarding the source of this new right, seven of the nine justices were in agreement that the intimacies of the marriage bed were beyond the control of the law.

The scope of this new right is still to be determined. But it is clear that the right extends beyond the narrow protection of the privacy of the marriage bed. The *Griswold* case itself involved prosecution of physicians who were active participants in a birth control clinic that gave physical examinations to women, advised them concerning the use of contraceptive materials and medicines and in addition supplied them with birth control devices. Some lower courts have found that this new right of privacy extends to *family, marriage and sex matters* generally, and most particularly to any procedure that falls within the area of birth control (7a).

The impact of the *Griswold* case had already made itself felt in an oblique way in the case law governing voluntary sterilization. In 1969, an indignant wife was denied a salpingectomy at public expense by a county hospital in California. She brought suit to have the operation provided as part of the county's free family planning and related health services for persons receiving public welfare. Her brief relied in substantial part on the *Griswold* case. The county defended on the ground that nontherapeutic sterilization was unlawful because it constituted unlawful maiming and was contrary to public policy.

The California District Court of Appeal never reached the

constitutional issue. It gave the California statute prohibiting maiming a narrow interpretation and was at a loss to find any public policy in opposition to birth control (8). One can only speculate concerning the influence that the *Griswold* case had in leading the court to find that California law did not stand in the way of a married woman's desire to be sterilized. However, it is bound to influence other courts. Since only a very few states have ever had an express statutory provision outlawing voluntary sterilization, most courts can avoid tangling with the *Griswold* decision by simply interpreting the law of their state as permitting voluntary sterilization. It is interesting to note that Connecticut, one of the few states having a statute expressly forbidding voluntary sterilization, repealed it in 1969.

STERILIZATION OF AN INCOMPETENT WARD BY COURT ORDER

Although the guardian of a mentally retarded ward has power to consent to a medical operation for the preservation of the latter's life or health, he lacks the inherent authority to consent to a non-therapeutic operation that will deprive his ward of the personal faculty of reproduction. At the very least he would have to apply for a court order expressly authorizing the operation.

A few lower courts have ordered such operations in the absence of a statute expressly authorizing them (9). In the one reported case taking this position, *In re Simpson,* an Ohio case, the mother of a mentally retarded girl with an illegitimate child sought the order as a guardian. In granting the order the judge gave three reasons for his action: the lack of room in the state's institutions for the mentally retarded; the welfare of society as a whole which would be endangered if the ward produced more illegitimate children who might in turn be mentally retarded and who would certainly not receive *the rudiments of maternal care;* and finally, the welfare of the mentally retarded girl herself. The court found that the operation would deliver her from a *lifetime of frustration and drudgery (10).*

Only the last reason could possibly justify the court to act without express statutory authorization. Theoretically, the court is the guardian of all wards within its jurisdiction. The appointed guardian acts only as an agent of the court. The court's power as guardian permits it only to take actions that benefit the ward. Therefore, in purporting to act for society's welfare, the court was exceeding its jurisdiction as guardian.

Two very recent cases confirm this conclusion. In fact, they go farther and support the conclusion that no court has the general power to order so radical an operation as sterilization for the benefit and well-being of an incompetent person without a grant of express statutory authority from the legislature. In a Texas case, the guardian of a mentally retarded woman with two illegitimate children sought to have her ward sterilized (11). The guardian, who was also the mother of the ward, was already taking care of the two children. Being in poor health she wished to avoid the physical, emotional and financial strain that would result if more children were born to the daughter. The court denied her petition on the ground that the fundamental legal rights of an incompetent person cannot be taken away, even when the guardian consents, without express statutory authority.

In the second case, the highest court of Kentucky refused to order the sterilization of a mentally retarded woman who had two illegitimate children, one of whom was also mentally retarded (12). The county health officer and the county medical society had been the parties seeking to obtain the court order for sterilization. In denying their petition the court said that there was no common law or statutory authority to grant them permission for such a serious operation. The Kentucky court expressly left open the question of whether a guardian could successfully petition for a sterilization order.

Arguably, the Kentucky court might grant such an order where the guardian makes out a case for sterilization on the ground that it would promote the general well-being of the ward. Shortly after deciding the sterilization case the Kentucky court decided the widely noted *Strunk* case in which it authorized a guardian, the mother of twin sons, one of whom was mentally retarded, to authorize a kidney transplant from her ward to his twin brother (13). The court found that the retarded brother would suffer a psychological trauma if his twin died. The court then proceeded to give a novel extension to the rule of substituted judgment. According to this rule the court may authorize action by the guardian if it concludes that the ward would have taken the same action if he were of sound mind. This doctrine is ordinarily limited to expenditures from the ward's estate, but here it was used to authorize an invasion of his bodily integrity. The Kentucky court split four to three on the transplant issue. There was a sharp dissent which argued that a mental defective would most likely suffer only transient psychological loss from the death of his twin brother.

Regardless of the desirability of the result in the *Strunk* case, or the merits of the majority's reasoning, it seems highly questionable to transfer the doctrine of substituted judgment to the sterilization area. First, it is highly problematical in some cases whether the ward's well-being is truly enhanced by the operation. There is at least one study that provides some evidence of resentment and disapproval on the part of feeble-minded persons who have been sterilized (14). Apart from the uncertainty concerning the benefit of the operation to the patient, there is the vexing question of whether such a personal decision should be left to whoever initiates the sterilization proceedings, even when it is the guardian.

On a question of morality that divides public opinion so sharply, can we be sure that the substituted judgment of the guardian reflects that of the patient? Studies have indicated that attitudes toward sterilization depend largely upon one's religious beliefs. Because of Roman Catholic opposition to sterilization, it is not likely that a Catholic guardian would ask for such an operation in the case of a Catholic ward. But suppose he does? Or suppose a non-Catholic guardian asks for the order in the case of a Catholic ward? Substituted judgment is of little help in this kind of case. Since the ward's mental disability is the reason for the operation, it is a logically impossible question to ask whether he would have consented to the operation if he had full use of his faculties.

To avoid this logical difficulty one might argue that the following question should be asked: "Would the ward, assuming he had full use of his mental faculties at some time, have agreed to be sterilized on the contingency that he would later become mentally retarded?" There is really no way of answering this very artificial question, particularly where the ward has been mentally retarded all his life. A court might understandably hold that a Roman Catholic ward would refuse such an operation had he full use of his faculties. But should the law assume such dedication to orthodox religious teachings? With regard to the use of contraceptives, there is considerable evidence that a number of Catholics are rejecting their Church's official teachings because they conflict with their own personal convictions.

Neither can we say with confidence that all mentally retarded non-Catholics would agree to such an operation if they could understand its implications. Opinion surveys show that a substantial fraction of non-Catholics oppose sterilization on moral grounds (15).

Finally, serious questions involving the separation of church and state would arise if the courts were to use religion as a basis for deciding when they would or would not take action affecting vital personal interests.

Because of the divided evidence concerning the beneficial effects of sterilization on the mentally retarded and the sharp disagreement in our society as to its ethical propriety, this is not a matter that should be left ultimately to the individual judge. Aside from the danger that personal predilections will enter into the judgment of the most objective of judges, there is the more fundamental question of whether such basic issues of value and policy should be decided by one man not directly accountable to the community. The legislature would appear to be the proper forum in which to consider and answer the questions raised.

The discussion thus far has assumed that sterilization of a mental defective is constitutional if done with the consent of the guardian. It has thus far dealt with the necessity of a statute authorizing this approach. The next question to be considered is whether such a statute would be constitutional or not. In the next section of this paper we will consider the constitutionality of compulsory sterilization laws. After considering that question, we will turn to laws authorizing sterilization with consent of the guardian to determine whether they stand in a more favorable constitutional light.

CONSTITUTIONALITY OF STERILIZATION STATUTES

Where a statute provides for the sterilization of persons convicted of certain crimes, there is a good chance that it will be held to violate the constitutional prohibition against cruel and unusual punishment. A few early cases held to the contrary, taking the position that cruelty requires some element of torture (16). However, the more forward-looking cases have held that the serious and usually irreversible consequences of the operation constitute cruel and unusual punishment (17).

Where sterilization is ordered for eugenic reasons, it is not vulnerable to consitutional challenge on this ground. The courts have generally held that the sterilization of non-criminal mental retardates and defectives does not constitute cruel and unusual punishment because the purpose of the operation is not punitive but promotive of the public welfare (18).

This public welfare justification would also seem appropriate where a criminal is sterilized for eugenic reasons rather than punitive ones. However, the courts are apt to scrutinize with great care the eugenic indications for sterilization in such cases in order to unearth underlying vindictive or punitive purposes. Most vulnerable would be laws providing for the sterilization of criminals on the ground that they have committed particular crimes, as is presently the case with the California law authorizing sterilization of persons convicted of carnal abuse of a child (19).

EQUAL PROTECTION OF THE LAWS

Compulsory sterilization laws have frequently been challenged on the ground that they single out a class of persons unfairly in violation of equal protection of the laws. Two kinds of discrimination have been attacked on this ground. The first kind arises from the fact that the large majority of state compulsory sterilization statutes apply only to inmates of state institutions for the mentally ill or retarded. It can be urged that these inmates are being discriminated against. An early New Jersey case struck down a sterilization statute on this ground, finding no rational basis to justify the difference in treatment (20). It found the statute particularly arbitrary because non-institutional defectives were more apt to have children than those in state institutions.

The Virginia statute challenged in *Buck v. Bell* was also limited to institutionalized defectives. It was challenged on this ground. Mr. Justice Holmes brushed aside the argument, belittling the significance of the equal protection clause of the federal constitution. He justified the difference in treatment on the ground that inmates of state institutions were more accessible to treatment than other defectives. He also pointed out that as more defectives were sterilized and released from state institutions, there would be more room to reach the others who otherwise might have children. The *Buck* case has not been greatly criticized on equal protection grounds and probably is still a controlling precedent on this point. However, as a matter of sound policy, there seems to be little reason to restrict eugenic sterilization only to inmates of state institutions.

The second kind of discrimination is based on the eugenic indications established by law to authorize compulsory sterilization. The inclusion of one class of defectives under a statute that does not include another class would be vulnerable on equal protection grounds unless there were some rational basis for distinguishing between the two.

Such an equal protection argument was successful in the *Skinner* case in which the Supreme Court of the United States struck down an Oklahoma statute which authorized compulsory sterilization of habitual criminals (21). A habitual criminal was defined as a person convicted two or more times for crimes *amounting to felonies involving moral turpitude.* The court found that this statute violated equal protection because recidivists who were guilty of felonies such as larceny by fraud could be sterilized under the statute but those only guilty of embezzlement, a misdemeanor, could not. The court could find no rational basis for such a distinction between two crimes of essentially the same nature and moral turpitude. Certainly, there was no eugenic justification that could justify this difference of treatment.

A number of state statutes include compulsory eugenic sterilization of epileptics as well as mental defectives, but others do not. The latter statutes might appear to be vulnerable under the

Skinner approach, but it is doubtful that they are. The difference in treatment that these statutes accord epileptics as compared with other defectives can probably be justified on both eugenic and social grounds. It should also be kept in mind that the *Skinner* case involved criminal behavior as a basis for sterilization. Two members of the Supreme Court noted in separate opinions that there was little eugenic justification for sterilizing felons generally in the first place. In view of the weak eugenic justification for the law, it is not surprising that the court was ready to find an invidious discrimination between different classes of criminals.

PROCEDURAL DUE PROCESS

Under the typical sterilization statute, only persons who suffer from certain mental illnesses or defects can be sterilized without their consent. In addition, some other statutory condition or conditions must invariably be met, requiring such findings as the following: that the individual to be sterilized is the *probable potential parent of socially inadequate offspring,* that the operation will be in his *best interests* as well as that of society, and that it will not endanger his health.

Procedural due process requires that the individual who is to be adversely affected by governmental action, or his legal representative, be given a fair and reasonable opportunity to challenge the state's contentions that the requisite conditions legally authorizing the state's actions in fact apply to his case. Procedural due process generally requires that the affected individual is permitted to contest them by cross-examination and the presentation of contrary evidence. The first compulsory sterilization statute passed in Indiana was eventually declared unconstitutional because it did not provide for a hearing and an opportunity to cross-examine the states's witnesses who had concluded that the operation was indicated (22).

The requisite hearing can be held before an administrator, such as the state secretary of health or his representative, or before a court. Where the hearing is before an administrator or his representative, it has been traditionally recognized that the administrative order should be subject to court review.

As a practical matter, in order for the individual to take full advantage of these procedural rights he should be represented by a lawyer, even when he is mentally competent. In other areas where important personal rights are at stake, it is now established that due process requires the appointment of an attorney at public expense in the case of an indigent (23).

Some compulsory sterilization statutes currently on the books are vulnerable to challenge on procedural due process grounds. A common flaw is the failure to provide free legal representation. Although the right to free legal services in this area may not be firmly established as a matter of constitutional law, no person should be denied access to basic legal safequards because of his economic status. Therefore, free legal services should be provided as a matter of sound social policy if not because of constitutional requirements.

SUBSTANTIVE DUE PROCESS

A final and most fundamental attack on compulsory sterilization statutes charges them with violating substantive due process. The legal concept of substantive due process rests on the proposition that in the case of a government truly subject to the rule of law, the safeguards of procedural due process are not enough. Under this rule the interests of the individual to act freely with regard to certain matters are found to be of much greater value than the interests of society in seeking to regulate his behavior. The state is constitutionally disabled from interfering with the individual's freedom unless it has a very good reason to do so. Careful draftsmanship can produce compulsory eugenic sterilization statutes that will withstand challenge on the other constitutional grounds, but objections grounded on substantive due process would invalidate all such efforts.

In *Buck v. Bell* the Supreme Court of the United States held that sterilization of mental defectives for eugenic reasons met the test of substantive due process. Speaking for the court, Mr. Justice Holmes found that the danger to society of being swamped with incompetents and the social problems created by them far outweighed the interests of mental defectives to propagate themselves and to be free of a relatively harmless surgical operation. At the time the *Buck* case was decided, prevailing constitutional doctrine held that a state could regulate the individual's behavior as long as the regulation bore a substantial and rational relation to public health, safety or welfare. Eugenic sterilization of mental retardates and other defectives readily met this rational basis test of constitutionality because of widely held scientific and social opinions of the time. Reliance on simple Mendelian genetic models led to the belief that some forms of mental illness and most forms of mental retardation were inherited. It was believed that sterilization of persons suffering from these defects would reduce their incidence in the total population. In addition, it was believed that the mentally ill and retarded were responsible for a disproportionate amount of crime and other social evils.

These opinions have since been subjected to considerable criticism. The etiology of most mental illness and mental retardation has been found to be a very complicated matter. The genetic code determining mental abilities and traits is far more complicated than the simple Mendelian model. The matter is further complicated by recognition that environmental factors also enter the picture. A number of commentators on *Buck v. Bell* have concluded that because of this complexity compulsory sterilization of mental retardates and defectives would have only a minimal impact on eliminating the incidence of these traits in the total population (24).

In 1937, the American Medical Association's Committee to Study Contraceptive Practices and Related Problems reported that *our present knowledge regarding human heredity is so limited that there appears to be very little scientific basis to justify limitation of conception for eugenic reason* (25). Earlier, the American Neurological Association's Committee for the Investigation of Eugenical Sterilization arrived at a similar conclusion (26). In 1960, Dr. Bernard L. Diamond, as a special consultant to the American Psychiatric Association in reporting on mental health legislation in British Columbia, stated that the concept of the inheritability of mental illness and mental deficiency is open to serious question regarding its scientific validity. He concluded that widespread use of sterilization procedures for the mentally ill or defective person is questionable (27).

Scientific opinion, however, is divided on this point. Reputable authorities on mental retardation in the United States and Denmark have recently concluded on the basis of statistical studies that the incidence of mental retardation could be diminished dramatically by sterilization of all defectives (28). These studies do not rely on a simple genetic explanation of mental retardation but automatically take into account environmental factors. This body of scientific opinion might be enough to satisfy the rational basis test, which does not require general agreement on the effectiveness or necessity of particular legislation.

However, in the intervening years since *Buck v. Bell,* the Supreme Court of the United States has held that more than the rational basis test is required to justify state action substantially infringing on certain fundamental freedoms. In such cases the state must have a *compelling subordinating interest* to justify its action (29). Not only must the state's interest be an important one, its regulations interfering with basic freedoms must be reasonably necessary to promote the state interest involved. So demanding is this new test that if the state interest can be reasonably attained by other means, then it must refrain from using ones detrimental to basic freedoms.

Compulsory sterilization laws infringe on two basic individual interests that are accorded this special constitutional protection. The first is the individual's interest in preserving his bodily integrity free from unwanted intrusion by the state (30). The second is his freedom to determine for himself such personal and intimate matters as the nature and quality of his sex life in marriage and the size of his family (31). In his concurring opinion in the *Griswold* case, Mr. Justice Goldberg expressly indicated that a policy of compulsory sterilization of married couples to control population growth would have to meet the *compelling, subordinating interest test* rather than the rational basis test (32).

There is serious doubt that the interest of the state in eugenic control of mental defectives can justify compulsory sterilization under the *compelling subordinating interest* test where scientific opinion is as divided as it is on the issue of heredity. However, the state need not rely soley on genetic justifications; it can advance environmental considerations. It is commonly assumed that mentally handicapped parents cannot provide the rudiments of an adequate home life for their children, who will consequently be stunted in their mental and social growth. The statutes of some states expressly provide that such environmental factors can support an order for sterilization. For instance, the Utah statute authorized sterilization of persons not capable of performing the duties of parenthood (33). Other state statutes impliedly recognize the environmental justifications. For instance, the North Carolina statute authorizes sterilization of mental defectives when it would advance the *public good* (34).

There is some psychiatric opinion that disputes the inability of feeble-minded parents to rear children adequately (35). On the basis of such opinion two justices of the Nebraska Supreme Court recently concluded that a eugenic sterilization statute could not be supported on either environmental or genetic grounds (36). However, the courts might accept the environmental argument more readily than the genetic one because it accords with their own common sense. Still they will be considerably reluctant to accept such a justification because of the dangerous precedent it would set.

If sterilization of mental defectives can be justified on environmental grounds then so, too, can the sterilization of other classes of persons who cannot provide an adequate environment for their children. Unmarried women who have given birth to one or more illegitimate children could be sterilized on the ground that they cannot provide a stable home life for their offspring. The law might become even more selective on the ground that the absence of a stable family is more important where children come from a deprived socio-economic background. Sterilization might therefore be made a condition that a woman on welfare would have to meet if she wished to retain custody of her illegitimate children. Such proposals have been introduced in state legislatures in recent years, although none have been enacted as yet (37).

The primary motive behind most of these proposed laws undoubtedly is reduction of the public welfare rolls. However, when challenged in court they would be defended on the ground of providing minimally desirable environments for children. The effectiveness of such laws in reducing various social problems, including mental retardation, would compare most favorably with strictly eugenical laws. It is generally recognized that environmental factors are very important in causing mental retardation (38). Some respected opinion suggests that these factors are even more important than the eugenic ones. Accordingly, a current precedent upholding the constitutionality of compulsory eugenic sterilization on environmental grounds strengthens the case for compulsory socioeconomic sterilization.

A New Jersey case pointed out these dangers as early as 1913. It said (39):

> There are other things besides physical or mental disease that may render persons undesirable citizens or might do so in the opinion of a majority of a prevailing legislature. Racial differences, for instance, might afford a basis for such an opinion in communities where the question is unfortunately a permanent and paranoid issue.

The potential for social oppression and invidious discrimination is so clear where compulsory sterilization on socioeconomic grounds is involved that one can reasonably expect our courts to strike down such legislation as unconstitutional. They would most likely hold that the state should deal with the demoralizing socioeconomic environmental factors of poverty, ignorance, etc. directly through appropriate social and economic legislation rather than finding deprived segments of society to comprise irredeemable misfits whose influence must be minimized by sterilization. But if compulsory sterilization for socio-economic reasons is unconstitutional, on what basis can compulsory eugenic sterilization be upheld when it is probably less effective in reducing the social ills against which both would be directed?

Two reasons come to mind. The first is the argument that the mentally defective person does not really suffer any serious deprivation when he is sterilized. Mr. Justice Holmes adopted this line of argument in *Buck vs. Bell.* Where low-grade mental defectives are concerned, the argument has some appeal. It is unlikely that such a person will feel his manhood or womanhood threatened, will miss the delights of parenthood, or will worry about the ethical and religious aspects of his situation. But these assumptions do not carry over to the feebleminded. As has already been noted, there is evidence of resentment on the part of the feebleminded who have been sterilized. Since there is also some psychiatric opinion indicating that mentally retarded persons can make good parents, the resentment is more than understandable.

Even in the case of the low-grade defectives, the lack of deprivation argument has unattractive features. It holds, in effect, that the mental defective does not have the same constitutional rights as the unmarried mother on public welfare. It moves in the direction of treating the defective as a means to an end rather than an end himself.

The second and more persuasive reason for treating mental defectives differently is that sterilization will often have the net effect of increasing their freedom rather than restricting it. As a practical matter, the release of retarded inmates from state institutions is frequently conditioned on sterilization (40). Similarly, state laws or administrators can make it very difficult or impossible for feebleminded persons to marry unless they first submit to sterilization (41).

Once again, however, we have the problem of the dangerous precedent. Will this justification grounded on freedom be, in fact, used to avoid a costly burden on the part of the state in caring for the low-grade defective? If such persons cannot understand the consequences of their actions nor reasonably regulate their sexual impulses, then should they be released from close supervision and placed in situations where others may take sexual advantage of them? Even more troubling is the case of the feebleminded person who state authorities believe is capable of a rewarding married life but is incapable of coping with parenthood. It is in this case that pressures leading to sterilization could possibly lead to lifelong feelings of acute deprivation.

What is most disturbing about this increased freedom argument is that the state can make it more persuasive by increasing the use of coercive practices. For instance, in a 1964 study, Professor Robert Levy found that under Minnesota's state guardianship law, officials were institutionalizing feebleminded persons who married and then releasing them only on condition that they submit to sterilization (42).

A very recent Nebraska case, *In re Cavitt* illustrates the current uncertainty surrounding the constitutionality of eugenic sterilization statutes (43). In that case a thirty-five year old woman with eight children and an I.Q. of seventy-one had been living with a man for fourteen years. After this relationship had broken up, she had difficulty in caring for her children and was admitted to a state home. Four years later the superintendent of the home sought a court order requiring her sterilization as a condition for release, as provided by state statute. The state statute was very broadly drawn. It only required findings that the patient to be sterilized was mentally deficient and was capable of bearing or begetting children. Once these findings were made, the board of examiners could order sterilization if they were of the opinion that it would be a proper condition for discharge. The trial judge found the statute unconstitutional but was reversed on appeal by the Nebraska Supreme Court.

Only three members of that court thought the statute was constitutional, while four did not. However, under Nebraska law a majority of five members of the state supreme court is needed to declare a statute unconstitutional. Most interesting is the fact that the three justices who upheld the law did not rely on the genetic argument. They adverted to the advances in medical science which had cast doubt on the inherited traits theory. Instead, they relied heavily on the environmental factors. They also took the position that the operation was not completely compulsory because the woman had the option of remaining in the home.

The dissenting judges rejected the notion that the operation was voluntary when the patient was faced with the alternative of life imprisonment. Two dissenting judges disputed not only the genetic justification for the law but its environmental basis as well. Speaking for them, Justice Smith said (44):

> I cannot imagine a causality more vague than that of heredity, environment, or both; and that very vagueness warns of menancing power over bodily integrity.

The other two dissenting judges found the statute unconstitutional because it lacked adequate standards. This position would strike down statutes allowing the full range of environmental factors to be taken into account in deciding whether to sterilize a mental defective. For example, that part of the North Carolina statute which permits sterilization of mental defectives for the *public good* could not stand under this test. These two judges were concerned about the potentialities for oppressive and discriminatory treatment under such a statute, pointing out (45):

> It is obvious that the board may arbitrarily, depending upon the particular thinking or philosophy of its members, exercise authority as to one inmate and not as to another.

These two judges were careful to point out, however, that they believed a statute specifying certain environmental considerations and factors could withstand a constitutional test. However, there is some question as to whether relevant environmental factors could be reduced to precise, workable guidelines.

STERILIZATION WITH THE CONSENT OF THE GUARDIAN

Consent of the guardian should not and probably does not render the operation voluntary from a constitutional point of view. A legal guardian has no power to waive the constitutional rights of his ward. Accordingly, even a limited statute authorizing sterilization with consent of the guardian must pass the various constitutional tests, particularly the ones of procedural and substantive due process. Such statutes probably stand on no better ground than ones that are completely compulsory.

With regard to procedural due process, there is still a problem of devising safequards to see that the interest of the ward is protected. The guardian will frequently have interests that might diverge from those of the ward. The guardian may be taking care of the ward's children and may be primarily seeking to gain his or her own peace of mind. For these reasons, a well drafted voluntary sterilization law should require a hearing with a guardian ad litem appointed to take a position adverse to that of the legal guardian. Some of the statutes currently authorizing sterilization with the consent of the guardian lack this safeguard.

Similarly, when we come to substantive due process limitations, a statute requiring the consent of the legal guardian must meet the same standards that are applicable to the more traditional compulsory sterilization statutes. Consequently, such laws are equally vulnerable to constitutional attack. The guardian's consent does not alter the basic intrusion into the patient's interests. The legal guardian has no more right or power than the state does to substitute his judgment for that of the mental defective. Indeed, since the legal guardian draws his power from statutes and from the courts, he is acting on behalf of the state. It is only proper that the same substantive due process standards apply in his case as in the case of the state. This conclusion is particularly compelling where the legally incompetent person has no natural guardian and a state official, such as the superintendent of a home for mental defectives, is appointed as legal guardian.

Although consent from even a natural guardian does not affect the constitutionality of a state's sterilization law, it makes it more socially acceptable than the strictly compulsory law. The opportunity for a law so qualified to be applied in a discriminatory and socially oppressive manner is limited because the consent of someone close to the ward would be required. For this reason the current administrative practice of states having compulsory sterilization statutes to limit their application to cases in which the natural guardian gives consent is to be welcomed.

CONCLUSION

The above discussion suggests that concern for the constitutional rights of mentally defective persons leads to a negative assessment of eugenic sterilization laws that do not require the informed consent of the individual to be sterilized. However, a very difficult problem arises in connection with feebleminded persons who can understand the consequences of sterilization up to a point and may want the peace of mind that the operation can give them. Such persons are apt to be very susceptible to strong pressures inducing them to undergo sterilization. Lawyers and psychiatrists working together should attempt to devise standards and procedures for determining how and when such persons can arrive at a decision that meets minimum standards of voluntariness.

REFERENCES

1. 274 U.S. 200, 207 (1927).
2. See notes 3 and 4.
3. Smith, Antecedent grounds of liability in the practice of surgery, Rocky Mt. L.Rev. 14:233, 277-78 (1940).
4. *Ibid*
5. In March, 1970 Congress established a Commission on Population Growth and the American Future to study in part "the various means . . . [to] achieve a population level properly suited for . . ." environmental, natural resources, and other needs [of the Nation]. Pub. L. No. 91-213.
6. Shaheen v. Knight, 11 Pa. C. & D. 2d 41 (1956).
7. Griswold v. Connecticut, 381 U.S. 479 (1965).

7a. E. g. Babbitz v. McCann, 310 F.Supp. 293 (E.D. Wisc. 1970). In the Babbitz case the district court went so far as to hold that the Griswold case implied a right that a woman has to abort an unquickened fetus.

8. Jessin v. County of Shasta, 79 Cal. Rptr. 359 (1969).
9. E.g. In re Simpson, 180 N.E.2d 206 (Prob.Ct. Ohio 1962). There are also unreported decisions by other lower courts in Ohio and Maryland granting such authorization.
10. 180 N.E.2d at 208.
11. Frazier v. Levi, 440 S.W.2d 393 (Ct.Civ.App.Tex. 1969).
12. Holmes v. Powers, 439 S.W.2d 579 (Ky. (1969).
13. Strunk v. Strunk, 445 S.W.2d 145 (Ky. (1969).
14. Sabagh & Edgerton, Sterilized mental defectives look at eugenic sterilization, Eugen Quart, 9:213, 1962.
15. Bass, Attitudes of parents of retarded children toward voluntary sterilization, Eugen Quart, 14:45-53, 1967.
16. E.g., State v. Feilen, 126 Pac. 75 (Wash. 1912).
17. E.g., Davis v. Berry, 216 Fed. 413 (S.D.Ia. 1914); Mickle v. Henricks, 262 Fed. 688 (D.Nev. 1918).
18. E.g., Clayton v. Board of Examiners, 234 N.W. 630 (Neb. 1931); Smith v. Command, 204 N.W. 140 (Mich. 1925).
19. Calif. Penal Code Sec. 645.
20. Smith v. Bd. of Examiners, 88 Atl. 963 (N.J. 1913).
21. Skinner v. Oklahoma, 316 U.S. 535 (1942).
22. William v. Smith, 131 N.E. 3 (1921).
23. Gideon v. Wainwright, 372 U.S. 335 (1963) established the constitutional right of all defendants accused of a crime to receive legal representation, requiring the state to provide such services in the case of an indigent.
24. E.g. Myerson, Certain medical and legal phases of eugenic sterilization, Yale. L.J. 52:618 (1943); Cook, eugenics or euthenics, Ill. L.Rev. 37:287 (1943).

25. Quoted in Ferster, Eliminating the unfit – Is sterilization the answer? Ohio St. J.J. 27:591 p. 599 (1966).
26. The Committee's conclusions are set forth in Ferster, supra note p. 599.
27. Diamond's conclusions are set forth in Ferster, supra note 25 at 600.
28. Reed & Reed, Mental Retardation: A Family Study. Phila., W. S. Saunders, (1965); Kemp, Genetic-Hygenic Experiences in Denmark in Recent Years, 1957 Eugenic Rev. 11.
29. The Supreme Court has gradually marked out a number of rights that can only be infringed by the state for grave and compelling reasons, starting with the First Amendment rights of speech, religion, assembly, etc.
30. The Supreme Court has a number of times remarked on the importance of maintaining bodily integrity to insure human dignity. E.g. Schmerber v. California, 384 U.S. 757 (1966); Skinner v. Oklahoma, 316 U.S. 535 (1942).
31. Griswold v. Connecticut, 381 U.S. 479 (1965).
32. 381 U.S. at 496-97 (concurring opinion).
33. Utah Code Ann.Sec. 64-10-7 (1953).
34. Gen. Stat. No. Car.Sec. 35-36 (1969 Cunn. Supp.).
35. Kanner, A Miniature Textbook of Feeblemindness. New York, Child Care Publications, 1949, pp. 4-5.
36. In re Cavitt 157 N.W.2d 171,179 (Neb. 1969) (dissenting opinion).
37. See Paul, The return of punitive sterilization proposals, 1968 Law & Society Review, 77.
38. E.g. Reed & Reed, Mental Retardation: A Family Study. Phila., W. S. Saunders, (1965).
39. Smith v. Bd. of Examiners, 88 Atl. 963, 966 (1913).
40. Nebraska until recently had such a law, but it was repealed shortly after four out of seven members of the Nebraska Supreme Court questioned its constitutionality. See discussion in text at notes 43-45.
41. E.g., So. Dak. Comp. Laws. Sec. 25-1-16 (1967).
42. Levy, Protecting the mentally retarded, Minn. L.Rev. 49:821, 832 (1965).
43. 157 N.W.2d 171 (Neb. 1969).
44. 157 N.W.2d at 180.
45. 157 N.W.2d at 184.

Chapter VII

THE POLITICS OF EUGENICS

ALYCE McL. C. GULLATTEE

THE subject of eugenics has so many emotional aspects attached to it right now that it is a matter of great concern to members of the black community, for all of the negative aspects that eugenics represent to blacks. This paper is entitled *The Politics of Eugenics* simply because it would appear that a great deal of talk about eugenics is wrapped up in or in some way related to politics.

Every twenty to twenty-five years it would seem, concomitant with the stresses resulting from an increasing world population and decreasing resources and space upon which the overflow will live, we Americans begin to reflect upon the problems of race and population. One convenient solution often suggested is some form of eugenics. Eugenics has been variously defined as a wedding of biology and politics with the proclaimed intentions of saving the human species. The group of humans worthy of redemption, however, remains ambiguously undefined. Those humans not suitable for perpetuation and procreation of the species are more easily identified by the prophets of ecological doom.

The eugenicist is neither a biologist nor an anthropologist, but rather a combination of the two, selecting out those traits and characteristics that tend to represent the negative and positive qualities of man. These qualities may be genetically or environmentally determined, but if the negative behavior or phenotype occurs with a degree of frequency and consistency, they are believed to be genetically bound. This genetic constituency will be described shortly.

The pertinent question at this point is, what is eugenics? The term derives from the Greek and means *well born.* Sir Francis Galton expanded the meaning to include the study of agencies under social control that may improve or impair future generations whether physically or mentally. The (biological) eugenicist ascribes to heredity the factors that determine what a man's physical or mental makeup will be. The environmentalist, on the other hand, claims that a host of social forces give shape to the human personality. He might argue further that the eugenicist may erroneously ascribe to hereditary determinism what are actually the effects of environmental phenomena. Under this erroneous assumption some eugenicists, race hygienists, and even some legislators, may conclude that an undesirable trait – for example, the hunchback – occurring in two or three successive generations is carried by the genes. An environmentalist or sociologist might have noted that the physical deformity was caused by years of cottonpicking or working in the coal mines where the body's attitude becomes a function of the uses of the body. Scientifically speaking, eugenic selection can only affect hereditary components.

At this point, perhaps, eugenics should be put into some sort of historical perspective. The theory of eugenics was first propounded by Sir Francis Galton in 1883. It sprang from the notion that defective persons procreate more rapidly and breed more readily than normals, the result being that society is flooded with inferior types and unproductive children. Early eugenicists proposed, in what was called model eugenic sterilization law, to eliminate through sterilization feeblemindedness, insanity, criminal tendencies, epilepsy, alcoholism, drug addiction, deafness, syphilis, tuberculosis, blindness, physical deformities and unproductive dependency such as pauperism, economic failure, and orphanism. The unabated propagation of human beings who were physically impaired, diseased and mentally deficient represented a danger as formidable as an invading horde of barbarians.

It must have been such a fear and justification that led, in 1895 the superintendent of the Wynnefield, Kansas State Home For The Feebled-Minded to order the castration of forty boys. Perhaps a similar explanation can be given for the sterilization of several hundred boys at the Indiana Reformatory even before laws were enacted approving this kind of procedure. Case studies done on the bloodlines of the Kallikaks, the Nams and the Jukes tended to reinforce prevailing beliefs about the inheritability of familial degeneracy and underscored the need for eugenic goals that could be enforced through legislation.

The first sterilization law was passed by the Pennsylvania legislature in 1905, but was vetoed by Governor Pennypacker. His humane remarks on the subject of eugenics and sensitive evaluation of the law and its broad general powers are worthwhile reading. The Pennsylvania law may be attributed to several precipitating causes, not the least of which were the first Negro convention in Philadelphia in 1830 – held to lend greater momentum to the abolitionist movement – an extraordinary occasion of an *inferior people* conducting a public forum; and after 1890 the steady mass influx of Slavic, Latin and Mediterranean immigrants to that state. Initially it was believed that integration and amalgamation of blacks and immigrant whites was possible, but the sheer number and strange cultural habits of the prolific *foreigner* created a threat. The demography, human ecology, political and social ideology of the State of Pennsylvania was not necessarily an isolated situation but rather represented and reflected a global overview of the attitudes in the United States.

Modern day advocates of eugenics and eugenic sterilization have been gaining momentum since the mid 1950's, but now the thrust for eugenic control is couched in terms of environmental pollution, zero population, the correlation between population and pollution, the quality of life, and whatever other euphemisms are available to mask the social issues of eugenics. It is readily conceivable in the present climate of things that eugenics legislation may be used to decelerate the birthrate among ethnic minorities, particularly the black minority. There are blacks, however, who would argue with some empirical justification that sterilization is a euphemism for black genocide.

A review of the literature brings to light an uncomfortable number of instances where to sterilize or not to sterilize was a matter of not too careful judicial reflection, dependent upon personal whim and bias rather than upon a careful appraisal of the facts and human rights. The fear has been expressed that the logic which led to the *Buck vs. Bell* decision might be extended beyond its present limited boundaries. In this case a child was declared to be an imbecile by a registered nurse. On the belief that three generations of imbeciles were enough the child's mother and the child were sterilized. Yet this child, it turned out, developed normally, and had already accomplished what was considered to be a normal education before dying at age six of measles. Someone had decided that because the mother was *a little bit abnormal,* according to whatever the normalcy standards, her child would, therefore, suffer from the same disorder.

Another case warrants attention here because of its importance for those interested in genetics: *Smith vs. Board of Examiners* (7). The following argument presented in 1913 would appear to be just as appropriate today:

> There are other things besides physical or mental disease that may render persons undesirable citizens or might do so in the opinion of a majority of a prevailing legislature. Racial differences, for instance, might afford a basis for such an opinion in communities where the question is unfortunately a permanent and paranoid issue.

According to some authorities, sterilization has proved a failure as a means of eliminating the *unfit,* but then some are more *fit* than others, to paraphrase the Orwellian line. Although very few sources give the racial breakdown of persons who have been sterilized, it goes without saying that where social factors are used as the basis for sterilization, a large percentage will be nonwhites, not because they are more often involved in violation of social imperatives, but rather because of the greater frequency of their apprehension and conviction compared to whites. Socioeconomic factors of blacks have variously been interpreted as cultural in origin resulting in the definition of norms of behavior that appear aberrant.

When eugenic laws were first proposed, a host of ills were recommended for elimination with no apparent concern for error in judgment or the credibility of medical and judicial evidence. The years, however, have revealed more information about genetically inherited and socially perpetuated disorders.

We now look at environmental determinants both *in utero* and *in vivo.* It is inconceivable that the white majority will destroy itself in the name of eugenics, and yet that is exactly what would result if chromosomal abnormalities and their phenotypic consequences of physical and mental inadequacies were controlled by sterilization. One-tenth of one percent of all births in the United States are Downs syndromes or twenty-one trisomy, and ninety percent of these are white. Errors in the number of chromosomes, the number of autosomes, and structural changes, such as in the arrangement of the chromosomes and chromosomal segments, occur more frequently in whites than in nonwhites. Eighty percent of the children who are born feebleminded come from normal parents. Klinefelter, Turner, Mongolism, Autosome 47, and Triple X syndromes are all genetically inherited characteristics that are found in far greater numbers in whites than in blacks. Inborn errors of metabolism also occur more frequently in whites than non-whites.

There are certain diseases that tend to present themselves only in certain ethnic groups, and after a number of intermarriages the familial traits tend to occur more frequently where bloodlines are obscure. This is true in lipid storage diseases, in Niemann-Pick's disease and Tay-Sachs' disease, all of which are found most often in Jewish stock resulting in cerebral compromise and idiocy. The blood abnormalities, such as sickle cell anemia, found almost exclusively in blacks; and Cooley's anemia, found in people of Greek, Italian or Arabic origin, may rarely have a component of mental deficiency associated with them.

There are other genetic disorders, such as Hand Schuller-Christian disease and Gaucher's disease, cystinosis, and phenylketonuria, which may affect the patient so severely as to warrant institutionalization. Knowledge of the inherited patterns of these disorders and others not cited here, such as teratogenic drugs and environmental conditions such as hypoxia, has forced a redefinition of the parameters and variables to justify eugenic sterilization.

Since 1964 there has been increasing opposition to sterilization on the grounds that scientific knowledge of hereditary factors in mental disability is not sufficient to warrant widespread use of sterilization, certainly not on an involuntary basis. Now that there is more information about genetic disorders, now when it appears that if one were to truly sterilize on the basis of genetic inheritance that a great many whites would have to be sterilized, this is just not going to come about.

Who then will benefit from sterilization laws? Society is still burdened with crime, poverty, mental illness, and mental retardation, with the latter two often being associated with poverty. Since the major cause for the recorded drop in sterilization procedures has been objections to it primarily on scientific grounds, more so than on moral and socio-political and economic grounds, the proponents began to find moral, socio-political and economic grounds to justify the continuation of eugenic practices. In fact, sterilization is still advocated as at least one of the solutions to the problems of crime, mental illness and retardation, all of which are considered to be secondary to poverty environments. Sterilization will save society from *poor parents* rather than *poor heredity,* or perhaps to be intellectually honest, we should say poor blacks or, again simply, blacks.

The three types of sterilization presently being suggested – voluntary sterilization of the mentally disabled on eugenic grounds, voluntary sterilization without any specific grounds, and involuntary sterilization of the unfit on grounds of social inadequacy – will affect in every instance nonwhites. Why? Because eugenic sterilization generally takes place in state institutions for the mentally disabled, and such institutions serve as incarceration sites for blacks if long-term psychiatric care is required since, historically, customs and/or lack of money has caused exclusion of blacks from private white sanatoriums. The penal institutions and welfare ward institutions are also sites of such practices, and sterilization may be recommended either by judicial directive or by psychiatric consultants. Blacks are also present in large numbers in such settings.

Liberalized abortion laws will provide the legitimate vehicle for *accidental sterilization,* as described by one physician to Dr. Julius Paul. The physician indicated that he frequently slipped and cut the fallopian tubes in a patient when he felt it was necessary, even without the consent of the patient, but when there were no nurses around. The laparoscope and the culdescope also make for easier illicit sterilizations.

Birth control uses such as oral contraceptives may result in sterilization procedures as has happened in cases where carcinoma *in situ* was found after the long use of birth control tablets. The patient is told that the proper medical treatment of choice would be for her to have her uterus taken out. As a result, there are a great many young women between the ages of twenty and twenty-six who are without uteri now, as a result of having been found to have carcinoma *in situ.*

In earlier times involuntary sterilization was urged by eugenicists as necessary to prevent inherited disorders. In states where such laws exist, they have sometimes been used for other than eugenic reasons, such as on grounds of social inadequacy. Now the claim is that not just the parent but the children should be sterilized because they will have the defects of the parents, who are too socially inadequate to fulfill the responsibilities of parenthood.

Bills for compulsory sterilization and unwed mothers have been seriously debated in a number of northern and southern states. Such laws could be used to coerce mothers who depend upon public funds for support to submit to sterilization. Poor blacks and, in this instance, poor whites would be seriously affected. What started in 1880 as a reflection of anxiety over the ever increasing numbers of *others* resulted in laws and political sanctions to control the unwanted in the last quarter of the nineteenth century; went ahead into the twentieth century to set the moral tone for the massive destruction of more than six million Jews in Europe in the 1940's; and reached the nonwhites in therapeutic eugenic wars in the 1950's and 1960's. Now these laws are being revived on the homefront, being rationalized as self-protection from the socially unfit.

It is contended here that such laws reflect a racist orientation and mentality. The unfit – variously defined as slow learners, minimal brain damaged, educable retardates, hyperkinetic children, emotionally disturbed, impulse-ridden proband and index case propositors – are being discovered by the hundreds in black, impoverished, economically deprived areas. Other tools of detection for cognitive learning difficulties are being used, such as ocular convergence and stereoscopic vision, problems with bifocal vision in the young preschool children, at a time when physiologically the ocular globe has not completed its anatomical growth.

What is there in our American society that has created a cultural ambience which would permit promiscuous and illegal sterilization to flourish under the guise of medically warranted eugenic procedure? I would like to suggest as causative factors those of stress pollution and the Brownian box phenomenon. The accelerated pace of American life has produced a similar increase in the need for instant resolution of problems, social and personal, which continue to grow in number and complexity. The psychological need to achieve and maintain a stable level of existence likewise forces people not only to struggle to attain status but to employ measures to see that no one takes it (and its prerogatives) from them. It is part of the racist thinking that the number of people competing for economic advantage must therefore be controlled. If blacks are involved, then the broader the range of sanctions that may be brought to bear to make them less or non competitive. If the number of blacks in the population can be reduced (as a complementary strategy to systematic economic disfranchisement), then their potential as competitors is reduced. But since blacks have little power to bring about effective change of the status quo, it is necessary for whites to examine carefully their racial attitudes on the present population crisis and on the nonwhite ethnic explosion before considering anything that has to do with the matter of eugenics.

It is difficult to see how a white person can achieve adulthood in this country without, in some way, being to whatever degree a racist. Racism is a particularly pernicious form of ethnocentric behavior in that the doctrine embraces without distinction the belief that racial and cultural traits are inherent and serves as justification for relegating blacks to a status which approaches servitude. Although we have a racially heterogeneous population in the United States, there is no other minority group, racial, religious or language, which faces the special situational experience of the black minority.

The interaction between blacks and whites at the primary level of emotion, rather than the secondary level of intellect, dictates the multi-faceted though often carefully hidden nature of racist behavior. This interaction is largely one of conflict. The aspect of conflict is therefore, potentially or actually, present in all black-white relationships. The nature of the conflict may be disjunctive, where there is a suspension of communication between the races, as presently is happening in some areas of the country. There is, of course, indirect conflict when individuals or groups do not deliberately impede the efforts of one another but nevertheless seek to attain their ends in ways that would consequently obstruct or impede the attainment of some end desired by the other group. Eugenics offers a way for this to come about. A working definition of white racism, therefore, would incorporate the idea of an attempt at all times and applying whatever modes of behavior, at both the conscious and unconscious levels, to maintain a position of superiority *vis a vis* ethnic minorities, most especially *vis a vis* blacks.

For the purposes of this paper, the point can be advanced that there are several kinds of racist behavior. There is, for example, the *active racist,* the one who has the power and uses it to reinforce his sense of racial preeminence. He is in a position to issue rewards and impose punishments as strategies for asserting his superiority. Eugenic sterilization laws are being used as such a strategy in places in the South to insure the permanence of white superiority. There is also the *latent racist,* generally the poor white whose status makes him ineffectual as an individual, although he is able to participate collectively in the power of the major ethnic group, whose reference of identification is the white authority structure of politics and economics, national and local. Presently, we are seeing in our country for the first time the largest number of educated poor whites to have reached middle-class status and, yet, being boxed in, are not able to move upward further, but are constantly being crowded from below, being placed therefore in a Brownian box. The poor white can move only horizontally, therefore, he must stop the numbers of people who can come into that box and compete with him in some way. His circumstances create anxiety for him. There is yet another, the *impotent racist,* who entertains notions of superiority because he can identify with the white man, but does not have the power to assert his appropriated sense of superiority. One need only take a look at some of the hyphenated Americans to see that they are placed in a position to carry out racist activities because they are looked upon as being different from the blacks.

At some point in a discussion of the active to impotent racist, one almost inevitably must deal with the individual who considers himself or is considered by others to be a liberal. In the present context, a liberal is defined as an individual who is relatively free of racial bias and the popular prejudices, who is willing to acknowledge the rights of others (even though the others are blacks and other ethnics) and who espouses, at least in the popular sense, equalitarianism. Such a liberal might live the ideal liberal behavior at the conscious level but at the subconscious level exhibit some racist attitudes. There are many liberals who unwittingly perpetuate the racist ideology, who because of some perceptual distortion fail to recognize their own racist mannerisms at the conscious level.

A total liberalism in all areas of the human enterprise, social, economic, religious, or political, is not and need not be an imperative condition. Blacks are not ultimately concerned about social equality and would settle for economic and political parity. Equality in these areas would provide the basis for a viable partnership, black and white. Participation in matters that have to do with eugenics and state eugenic laws is an absolute must for a viable coexistence between the races.

In many instances, people are not aware of their participation in the destruction of others. Americans are well aware of the fact that we, in this country, are just as responsible for the death of six million Jews as were the Germans who destroyed them because we turned them away from our shores. This might be regarded as a method of eugenics because it kept a relatively pure strain of Americans in this country and did not allow an alien people harbor and sanctuary in our country.

This is the direction in which we are going now and I think this is what is being felt by a great many people who are frightened by what it is that eugenics represents right now in the country. It is masked in other terms, of course, but the direction is the same. We convicted the Nuremberg principals for their part in compulsory eugenic sterilization, but we also have sterilized people. We are also culpable. It is only a matter of degree.

We are about to do the same thing now. Therapeutic wars cannot go on forever as a way to control population, such as, for example, the destruction in Viet Nam, the Near East and the Far East. The practice of searching one's soul, so long the private business of the individual, has suddenly loomed forth as a needed form of political and cultural sharing. For the effects of that which emerges from the psyche at the reality level touches the lives of others, vitally so, and reflects the real, the facade, the subliminally suppressed, and the repressed aspects of our personalities. The commentary, the rhetoric and open dialogue today on that concept of social ills called racism, which we can call social eugenics in this instance, are surface pools of ill-defined fluidity which hopefully can be easily evaporated by momentary gusts of a warm sensitive humanity. This vortical hate, centripetal and centrifugal movement, keeps in perpetual motion the dynamics of the unbearable activities of racism. Our vision, the vision of the black, of the contemporary American scene fills us with tested pessimism, with doubt and with disbelief, with extraordinary disenchantment with the American dream, with experience-taught cynicism, with suspicion, which is an understatement of the liberal rhetoric, with dark images of things to come. We regard with concern, indeed alarm, programs currently being proposed, such as Hutschnecker's proposal of testing six year olds for criminal tendencies, and the so-called Child Development Centers where children are being tested for retardation on the basis of the way in which they see. Presently in our cities we view with grave misgivings the irresponsible, immoral, illegal conduct of public officials, whether the style be that of equivocation or double-talk.

If we read the program designs of our President and Vice-President, we are led to conclude that the goals are wholely antithetical to the interests and welfare of nonwhite people and some ethnic minorities in this country. The continued violence practiced upon these people with governmental sanction, both local and national, suggests the growing institutionalized character of ethnic genocide, under the guise of improving the racial stock, by eliminating or reducing the transmission of genetic-linked diseases.

The systematic control of so-called inferior populations can be achieved. It is imperative that before any program of eugenics be initiated, those traits regarded as deleterious be unequivocally identified as biologically transmissable. This is to say that hereditary factors must not be confused with environmental factors. The idea of forced birth control, coercive birth control, or compulsory birth control, is morally repugnant, representing as it does not only an invasion of privacy, but an invasion of basic constitutional rights. The laws, no matter how they are passed, ultimately will only be as effective as the protection they give to the least of its constituency. We will all live together or we will all die together.

REFERENCES

1. Cromwell, James: The Negro in American History, Washington, American Negro Academy, 1914.
2. Danes, Betty S.: Genetic counseling, Medical World News, November 6, 1970, pp. 35-41.
3. Ferster, E. Z.: Eliminating the unfit: Is sterilization the answer? Ohio State Law Journal, 27:591-631, 1966.
4. Kindregan, Charles P.: Sixty years of compulsory eugenic sterilization: Three generations of imbeciles and the Constitution of the United States, Chicago-Kent Law Review, 43:123-143, 1966.
5. Paul, Julius: Population "quality" and fitness for parenthood in the light of state eugenic sterilization experiences, 1907-1966. Population Studies, 21:295-299, 1967.
6. Paul Julius: The return of punitive sterilization proposals. Law and Society, 3(1), 1968.
7. Smith vs. Board of Examiners, 85 N.J.L. 46, 53, 88 A. 963, 966 (1913).
8. Stoddard, L.: The Rising Tide of Color, New York, Charles Scribner's Son, 1920.
9. Stoddard, L.: The Revolt Against Civilization, New York, Charles Schribner's Son, 1922.

Chapter VIII

VOLUNTARY EUGENIC STERILIZATION

MEDORA BASS

THE PURPOSE

THE purpose of this paper is to assemble evidence to point out the urgent need to review the practice and principles of voluntary eugenic sterilization and to urge that all methods of birth prevention including sterilization and abortion be made readily available to all persons who need and want them, including the mentally retarded and the handicapped.

THE NEED FOR REEVALUATION

A reevaluation of eugenic sterilization is important because of the following developments:

> At present 95% of the mentally retarded are living in the community. When the eugenic sterilization statutes were being passed, the majority lived in institutions, and compulsory sterilization served as a means of returning them to the community, instead of being institutionalized for the sole purpose of preventing reproduction.
>
> More effective contraceptives have been developed: *the pill,* the new copper intrauterine devices and the injectables. There is accumulating evidence that many of the retarded can use contraceptives effectively (Fujita et al, 1967). Planned Parenthood Institute of Southeastern Pennsylvania reported taht there were several retarded individuals using each of these methods effectively, 1971.
>
> Many parents of retarded children now believe marriage can be supportive and stabilizing to the higher grade mentally retarded (Bass, 1964; Goodman et al, 1971). Other parents are realizing the lack of heterosexual relationships and the segregation of the sexes may foster homosexuality. Some are working to set up coeducational residences

in the community, and are concerned about possible unwanted babies. Some parents are facing the broad problem of the prevention of hereditary types of retardation and have recently established a committee on prevention.

Recent studies in genetics have made a new look at sterilization timely. The longitudinal study of Reed and Reed (1965) included 82,000 relatives of 289 retarded inmates of an institution and covered as many as seven generations. They found that when both parents were retarded 39.5% of the children were retarded; with one retarded parent 11.2% were retarded and with both parents normal 0.9%; when both parents were normal and had normal siblings, the percentage of retarded offspring was reduced to .05%. The risk of having a retarded child if both parents are retarded is forty to eighty times as great as it is if both parents are normal. They also found from another unselected sample of retarded children that 48.3% had one or both parents retarded. Benda et al (1963) studied retardates of the cultural-familial type in the Fernald School and found 48.8% had one or both parents retarded. Reed and Reed (1965) concluded from their data that, regardless of the relative influence of nature or nurture, mental retardation could be reduced 17% in the first generation if the retarded did not reproduce. This figure may appear less than indicated by the data above, due to the fact that only 47% of the females and 24% of the males in their unselected sample ever reproduced. The number reproducing may increase with the present trend of *normalization* of life for mentally retarded individuals.

Recent experience in Denmark has supported this conclusion. Kemp (1957) estimated that the rate of mental retardation had been reduced 50% between 1925 and 1950 by a broad program of sex education in the schools, genetic counseling, readily available birth control, voluntary sterilization and abortion.

The recent census data from Omaha, Nebraska, indicate that a similar decrease may be beginning in this country because of the significant drop in the birth rate of the lower socioeconomic groups which produce a disproportionately large number of damaged and retarded infants. The decrease for four economic groups from highest income to lowest was: 0, 17%, 24% and 37% for the poverty group. In one Negro ghetto the decrease was 67%. Increased use of contraception among the high risk unmarried girls and the high risk older women may have also contributed to a decrease in the rate of mental retardation (Wolfensberger, 1971).

In the last twenty-five years in Pennsylvania the rate of

institutionalized retardates has increased from eighty-three per 100,000 to 129, and the waiting list has increased in the last twenty years from 924 to 5267 (Public Welfare Report, 1969). During this period birth control for the mentally retarded was practically unavailable.

THE DEFINITION OF EUGENIC

The word *eugenic* comes from the Greek *well born.* The dictionary defines it as relating to healthy children or improving the human race. Positive eugenics, or improving the race, meets with almost superhuman difficulties, and there is a wide belief that this meaning should be dropped. Negative eugenics arose in connection with Darwin's theory of evolution and the survival of the fittest. The Danish refer to negative eugenics as medical genetics and genetic hygiene. Kemp (1957) described eugenics, or medical hygiene, as resting definitely on the principle of voluntariness, undertaken exclusively at the desire of the person concerned. "It is sometimes stated that the main purpose of genetic hygiene (medical genetics) is to spare the community expense. This is, of course, a great mistake. Negative eugenic measures are of an entirely medical character aiming at preventing disease and misfortune" (*Ibid.* p. 12).

A pediatrician, Zellweger (1967), writing on genetic counseling, refers to eugenics as *preventive* and states that it does not really alter the gene pool of mankind, but its goal is to prevent individual cases.

Birnbaum (1961) defined eugenic sterilization as being performed because it is believed the mental disorder is hereditary and may severely handicap any future offspring, and because it is believed the mentally disordered person cannot properly care for or rear any future offspring. *Eugenic* in this paper will cover both of these concepts and will focus on the healthy child.

Birnbaum (1961) also refers to *eugenic institutional sterilization* as applying to those segregated in institutions for the sole purpose of preventing reproduction. An attempt to get data on the present numbers ran into much difficulty. However, there is some evidence that the total is considerable due to the peak admissions of adolescent girls and young women.

HEREDITY AND ENVIRONMENT

The long controversy between heredity and environment has held up action on voluntary birth prevention for the retarded. Nature-nurture are so interrelated that it is doubtful the exact contribution of each factor will ever be accurately measured. Whether heredity or environment is responsible for mental retardation is not the most important factor in the consideration of marriage and parenthood, because the retardate is seldom an adequate parent whether or not the child is normal. A child born to a retarded couple has a small probability of growing up to be normal unless intensive programs of rehabilitation for child and mother are made available. A Danish follow-up study of 352 children of retarded parents showed only 5% of the children normal emotionally and mentally *(op. cit.)*.

Other studies (Charles, 1957; Kennedy, 1948) in this country have shown a higher rate of normalcy. This points to the need for more research in the child rearing practices of the mentally retarded. Mitchell (1974) found that a social worker could do little to improve child care; Mickelson (1949) investigated the possibility of helping the retarded give better care to their children and found that sterilization was the most important factor in the satisfaction of child care, rather than supervision of institutionalization.

Heber (1970) in his study of *slum dwelling* children found the high prevalence of retardation associated with slums is not randomly distributed but is "strikingly concentrated within individual families and can be identified on the basis of maternal intelligence." Rehabilitation programs for the mothers and intensive stimulation of the infants, beginning soon after birth, have given some evidence that the previously observed decline in the children's IQ may be prevented.

Great progress has been made in the control of infectious diseases, and patients and physicians are becoming more concerned with genetic disorders which may indicate permanent contraception.

Acceptance of sterilization by Catholics appears to be increasing.

> Hall (1965) found 69% of one large sample of sterilized women – excluding Puerto Ricans whose preferred method of birth control is sterilization – were Catholic (1965).
>
> Hall (1965) reported gross inequity in hospital practice between the availability of sterilization and abortion for ward and private patients. The average ward patient had 6.1 children, the private 3.8 before they were offered sterilization.
>
> Some institutions for the retarded are reviewing their stand on birth control. Recently, a Catholic psychiatrist was instrumental in introducing sex education and birth control services in a state hospital in Iowa.

A reevaluation is most important for psychiatrists in many states, but even more so for Pennsylvania psychiatrists because the Pennsylvania practice is one of the most irrational in the country. Pennsylvania has a law that *the weak-minded* may not marry; Pennsylvania also had until very recently an institution especially for *feebleminded females of child-bearing age.* At the same time that the state attempted to prevent marriage and child-bearing, the Department of Welfare prohibited any social worker to mention birth control to a woman on welfare, and many of the retarded depended on welfare for their medical services.

This schizophrenic policy of attempting to prevent reproduction by prohibiting marriage and promoting *institutional sterilization* on one hand and denying birth control on the other hand may have contributed to the 50 percent increase in the proportion of Pennsylvania citizens in institutions and increasing by 500 percent the number on the waiting list.

A recent effort to liberalize the law so that the *weak-minded* might marry, if sterilized, was unsuccessful. Physicians might use their influence to abolish this archaic law.

BRIEF HISTORY

Sterilizations may be categorized as punitive, religious, therapeutic, eugenic, and contraceptive. Historically, sterilization was first associated with the castration of the enemy as punishment and to provide eunuchs. Later, emphasis was placed on the religious aspects. Castration was practiced among some of the early Christian cults and was used until 1884 to preserve the boyish voices of those in the Sistine choir (Fletcher, 1960). Therapeutic sterilization appeared later with the development of surgery. After Darwin and Galton, eugenic sterilization was used quite extensively in the wake of the theory of the survival of the fittest and the belief that criminality and degeneracy were due to heredity. It was believed that the unfit could be eliminated by sterilization. Approximately 3,000 operations were performed a year in the 1930's as compared to less than five hundred in the 60's (Wood, 1967). This decrease occurred without any change in the law. Birnbaum *(op. cit.)* asks the reasons for this: "Is it due to a lack of belief in heredity; a change in administration policy; better facilities, more adequate welfare programs, fear it was being abused or fear of a law suit?" He believes that answers to these questions might lead to a more rational and ethical policy regarding sterilization.

In 1931 sterilization was practiced by Steinach to rejuvenate men. In the 1950's, the contraceptive aspect of sterilization gained prominence in Puerto Rico as a permanent method of birth control and has been gaining acceptance in this country, especially since the public hearings on the dangers of the pill. In Puerto Rico, a recent study found 32% of married women between the ages of twenty and forty had been sterilized. In the United States, the number increased from 8% of all married white women in 1960 to 11% in 1965. The percentage of husbands sterilized had increased from 1% in 1955 to 3% in 1965 (Presser, 1969). It is estimated that approximately 1,000,000 sterilizations were performed in 1971 for contraceptive purposes, compared to 100,000 a year in the previous decade.

At the risk of oversimplifying, it may be said that historically sterilization originated as punitive, changed to religious, developed to therapeutic, was later used as eugenic and now has increased in use as contraceptive to such an extent that this category of voluntary sterilization for normal persons far outnumbers all others. Sterilization has almost entirely ceased to be punitive and has become a privilege which increases freedom of choice.

Some attempts have been made recently to require sterilization of women on welfare who have had more than two illegitimate children (Paul, 1968) but these moves have been defeated in every case. Legislation was introduced in Tennessee in 1971 to withhold welfare payments to mothers with illegitimate children unless they agreed to be sterilized. This was defeated and another bill passed making funds available for free sterilizations.

REPORTS OF COMMITTEES APPOINTED TO STUDY EUGENIC STERILIZATION

This investigation has revealed no evidence of any report of a medical or social welfare committee appointed to study the problem of sterilization which has not recognized the advisability of voluntary sterilization under certain circumstances and with proper safeguards.

Myerson (1936), reporting for the committee appointed by the American Neurological Society, wrote in 1936:

> There need be no hesitation in recommending sterilization in the case of feeble-mindedness, though it need not, of course, be urged in those conditions which are of definite environmental origin. Though we hesitate to stress purely social necessity for sterilization, it is obvious that in the case of the feeble-minded, there may be a social as well as biological situation of importance. Certain of the feebleminded can only, under the most favorable conditions, care for themselves, and a family of children may prove an overwhelming burden.

The White House Conference in 1930 reported " . . . there are no surgical, legal, or humanitarian obstacles to the extensive practice of selective sterilization" (1930).

In England, the Department Committee on Sterilization (1934), sometimes referred to as the Brock Committee, appointed in 1934, unanimously recommended voluntary sterilization after studying evidence supplied by sixty experts in the field. They recommended that voluntary sterilization be legalized and the *right to sterilization be extended to all persons whose family history gives reasonable grounds for believing that they may transmit mental or physical defect.*

A Swedish study of 225 women who had been sterilized came to a similar conclusion: *Too much weight may often be given to the risk of psychological complications. Indeed, preoccupation with risk may cause the operations to be denied to those handicapped people who need it the most* (Lancet Editor, 1962).

Studies of the psychological effects and the satisfaction of males and females has been summarized by Presser (1970). The satisfaction rates for males range from 68% to 100%, with the average for those in the U.S. 95%. For females the range is 78% to 98%, with the average for the U.S. 95%.

Another Swedish study (Weintraub, 1951) found that the law was seldom used without the consent of the individual and that almost all persons who were able to live in the community were able to understand the purposes and consequences of the operation and did give their consent.

The White House Conference on Mental Retardation in 1963 stressed the need of birth prevention and the possibility of reducing the rate of retardation (Dunn, 1963).

The President's Task Force, (1970) on the Mentally Handicapped made as its first substantive recommendation that information on birth control, birth control services, voluntary sterilization, and abortion be made readily available to all persons.

A poll of physicians in 1963 showed that 79% approved of voluntary sterilization in cases where mental deficiency was hereditary; 66% approved when there were too many children in relation to income (*New Medica Materia,* 1962).

LEGAL STATUS

Concern for the legal risks involved for the physician renders it difficult for many patients to obtain sterilization. This investigator found thirty-five percent of one sample of non-retarded persons applying to the Association for Voluntary Sterilization (AVS) for sterilization had previously requested the operation and had been refused.

Voluntary sterilization is legal in all fifty states except Utah where it is limited to reasons of *medical necessity;* however, Utah recognized the right of the retarded to childless marriage (Allen, 1968). There are no legal restrictions on the purposes for which the operation may be performed.

Georgia, North Carolina and Virginia have passed enabling laws which include the consent of spouse. North Carolina's law makes it possible for an abandoned wife to have the operation without the husband's consent. However, in practice, doctors and hospitals in other states are extremely reluctant to perform a sterilization without the consent of the spouse.

The Virginia law may be interpreted as requiring that all operations be performed in a hospital. The law also requires a thirty day waiting period which precludes salpingectomy for the mother with more children than she wants who comes to the hospital only for delivery. These restrictions point out the necessity of extreme care in passing the enabling law which may prove too restrictive.

There is no record of any physician losing a law suit for nonnegligent performance of voluntary sterilization when the proper signed consent form had been obtained. Twenty-seven states have laws applying to the sterilization of a person before release from an institution. Nebraska, in 1969, repealed this law. Many professionals in the field, including those working with the AVS, believe all compulsory laws should be repealed. It has been reported that individuals were seldom sterilized against their will. However, there have been reports of persons being sterilized without their knowledge. If AVS had the funds, it would investigate this because of their policy of requiring an informed consent before a referral is made.

The whole area of the validity of consent of the retarded needs study as does the right of a parent to have a retarded child sterilized. One mother of twin sons with IQs in the fifties spent three years trying to find a doctor who would sterilize her sons. She applied to mental health-mental retardation officials, to lawyers, judges, district attorneys. She finally found a sympathetic doctor who was a friend of their family physician. After the operation was begun, the nurse walked out for religious reasons, and the mother had to take over the duties of nurse. Society in its attempt to protect children from unscrupulous parents is preventing conscientious parents from doing what they believe is in the best interests of their child. This points to the need of study and recommendations by a medico-legal group.

A law was introduced in California which would have required hospitals to apply the same policies to sterilization as to any other surgical procedure; that there should be no special restrictions not applying to other operations. This bill passed the Senate but was tabled by the House of Representatives. Some hosptials have already adopted this policy.

Davis (1970), reported that it was important for doctors to know that the legal dangers of sterilizing a mentally competent person are mostly non-existent at this point. However, the AMA (1965) advised:

> The physician should be extremely cautious about sterilizing an incompetent person on consent of guardian. He would be well advised to determine whether consent is legally recognized by local statute. If there is no law, it is inadvisable for him to take responsibility himself, regardless of any concurring opinion he may obtain from medical consultants.

ECONOMIC ASPECTS

Medicaid funds cover sterilization in forty-seven states and the District of Columbia for medical necessity and in thirty-three states for socioeconomic reasons. Sterilization is covered specifically in forty-five Blue Cross and forty-three Blue Shield Plans for medical necessity and in forty-one and thirty-seven plans, respectively, for socioeconomic reasons (AVS, 1972). Private insurance companies vary from company to company and with individual policies. Every effort should be made to have these companies liberalize their policies so as to cover sterilization and abortion. The Medical Societies with their prestige could be most effective in bringing pressure on insurance companies to include sterilization.

MORAL AND RELIGIOUS ASPECTS

The moral aspects are based on the benefits of voluntary sterilization to the individual, to the unborn child, to the parents and to society. According to Fletcher (1960), moral status is based on freedom to make a choice and have the facts to choose between alternatives.

At the 1971 Symposium on Choices of our Conscience

sponsored by the Kennedy Foundation, Professor Nicholas Hobbs of Vanderbilt University said:

> The human rights of the retarded are in no way distinct from the rights of other individuals . . . provision should be made within the institution for the exercise of sexuality. For example, it should be possible for institutionalized retardates to marry or live in unions that enrich their lives and build stable and rewarding relationships.

Hobbs maintained that they should not have the right to unlimited reproduction, and he suggested that perhaps the rights of all persons to unlimited reproduction should be restricted. A Jesuit professor of ethics from Loyola University strongly urged that if the mentally retarded individual is unable to sustain a *mothering* relationship and is unable to understand a conjugal relationship the person's need for affection should be *channeled away from genital expression.* But, if it is impossible to *protect the retarded woman against such relationships, then – when other forms of protection will not succeed or would be oppressively inhuman – contraception or sterilization should be used to prevent the prospective harm . . .* This advice, in sharp contrast to the traditional Roman Catholic prescription of birth control may indicate that the Catholics may be changing their view on the morality of sterilization.

Warren R. Johnson (1965), after years of experience in a program for the retarded, wrote that the retarded should have the rights of other human beings " . . . but is parenthood the automatic right of all human beings? So often we neglect to raise the crucial question: What about the rights of the offspring of the retarded person and the rights of society?"

Castration is mentioned without approbation in the Old Testament. Jesus said, "Some have made themselves eunuchs for the kingdom of Heaven's sake" (Matt. 19:12). The major Protestant denominations have approved of voluntary sterilization for valid reasons, as have some rabbis.

BENEFIT TO THE RETARDED INDIVIDUAL

Birth prevention for the retarded is based on the principles of enabling them to live as full and normal a life as possible and protecting them from responsibilities and stresses beyond their abilities. Many retarded couples are able to maintain themselves in the community and can somtimes cope with one child. However, the incidence of mental illness is much higher than in the general population (Masland, Sarason, Gladwin, 1959). Pollock (1945) found seven times the number of admissions to a state hospital for mentally retarded persons. Penrose (1934) wrote that the added responsibility of a child often overburdened the capacity of the retarded individual, added to his feeling of inadequacy and jeopardized his ability to live outside an institution. Pavenstedt (1965) reported the very low lower class mothers were impulsive and inconsistent in their child rearing practices, and one cannot expect them to find value in their children until they are able to sustain feelings of self-confidence and self-respect.

We tend to believe the myth that having children provides only joy and happiness and overlook the disappointments and frustrations. Menninger (1942) wrote: "It is a reductio ad absurdum to attempt to force this responsibility on people in the face of real obstacles." All too often, lack of sex education and birth control services do force the retarded into situations where they will experience nothing but failure and tragedy.

There is some evidence that the sterilized retarded individual may make a better adjustment after release from the institution than the nonsterilized (Johnson, 1946; Shafter, 1957). Young girls are often returned to the institution because of pregnancy.

We lack studies on the motivation of the retarded for parenthood. Sarason (1953) wrote: "The defective mother does not plan to have her child, probably experiences its presence as an unnecessary annoyance, does not possess adequate knowledge or receive guidance in child care, and is probably more concerned with her own than the child's needs."

Retarded individuals are likely to function in the lowest socioeconomic class due to the competition in the labor market. This competition will become more difficult as more of the unskilled jobs are taken over by machines. With appropriate guidance, many retarded young adults realize that it is in their own best interest not to have children. Retarded couples are applying to AVS for sterilization after having been refused help by doctors and hospitals. In one case, a retarded mother had two retarded children and begged to have her tubes tied. She was refused and had three more children before she was offered birth control by a sympathetic social worker. She had two more children before she was given the operation which she had wanted for so long. Now seven retarded children are in institutions because this mother was refused help by the medical profession and the hospital.

THE UNBORN CHILD AND INADEQUATE PARENTS

Forty years ago, the White House Conference on Children and Youth resolved that no child should be born in America without the birthright of a sound mind in a sound body and who had not been born under proper conditions. Recently, I was asked to speak to the staff of a sheltered workshop because two girls with IQs in the forties had become pregnant in one month and both the grandmothers were retarded. This leads one to ask: How many retarded children are in foster homes awaiting adoption?

Coakley (1959) wrote:

> I think we need to know much more about the ability of the mentally deficient to offer adequate care to children. We are having a number of unmarried mothers referred to us, some of them with IQs in the forties. Another group we are concerned about is that of cases referred to us for service on a protective basis, children with one or both parents defective, many of these border on neglect.

A study of a small sample of retarded married males found four out of five unable to cope with children, and these became the responsibility of others (Peck et al, 1965).

Mitchell *(op. cit.)* and Mickelson *(op. cit.)* Charles *(op. cit.),* Jastak et al (1963), Kennedy *(op. cit),* Kratter (1958), Shaw and Wright (1960), and Sheridan (1956) studied retarded families and concluded many borderline defectives make fairly satisfactory housekeepers if their circumstances are familiar and they do not have many children.

Mattinson (1971) reported at the recent Conference on Human Sexuality and the Mentally Retarded, sponsored by the National Institutes of Health in Hot Springs, Arkansas, on a study of thirty-two retarded couples with IQs ranging from thirty eight to ninety-three. Nineteen percent were on welfare, 25% required one visit a week from a social worker, 25% were unknown to a social agency, one-third were steadily employed. She concluded that one mother with an IQ of forty-eight, another of forty-one could provide adequate care for preschool children. Six of the forty children had been removed because of neglect.

Heber (1970) found in his study of slum children that mothers with IQ's less than eighty, although accounting for less than half the total group of mothers, accounted for almost four-fifths of the children with IQ's below eighty. His data also showed a greater number of children in families where father and mother tested below seventy. "The adverse consequences of this differential in reproductive activity are, of course, of great social concern irrespective of one's views of the etiology of the intellectual deficiency in the parents." (Heber, 1970).

Kagan (1969) has found that a child lives and develops through satisfying interaction, mutuality, stimulation and pleasure. This pleasurable interaction and intellectual stimulation is apt to be missing, especially as the child of the retarded parent grows older.

Malnutrition, illegitimacy, divorce and higher death rates are all factors to be considered in evaluating the child rearing practices of mentally retarded parents.

FOR THE BENEFIT OF PARENTS

Too little attention has been given to the dread and frustration of parents of the retarded and the relief they may experience when marriage is made possible and the threat of an unwanted baby has been removed. Parents of the retarded have to live every day with one of life's greatest burdens. Added to the tragic disappointment, guilt, constant supervision and frustration, is the fear of their daughter becoming pregnant or their son being accused of paternity. No matter how much training or supervision parents give, there is always the possibility of sexual exploitation. The constant supervision is galling to the mother and the child and may lead to greater overprotection and dependency and feelings of rejection and inferiority in the child.

Many parents sincerely believe that marriage would be beneficial to their young. Hammar et al (1967) found 65 percent of parents thought their young would marry, and 70 percent considered the prevention of reproduction to be imperative for the health and safety of the offspring.

Research (Bass, 1967) on attitudes of parents of retarded toward voluntary sterilization revealed that almost twice as many parents approved as disapproved, 60 percent versus 33 percent. Many of these parents believed it would be in the best interests of their child to be sterilized despite the erroneous belief that it was illegal, castrating, and forbidden by their church. These parents conflicts were exacerbated by the taboos and lack of information in the area of birth control and sterilization.

FROM THE POINT OF VIEW OF THE COMMUNITY

Decisions regarding the retarded must be considered in the total context of the child in the family and the family in the community. Poverty has often been given as the cause of 75 percent of retardation. The reverse is true also; retardation may lead to deprived multi-problem families, and these multi-problem families consume 60 percent of our welfare funds and services.

Our society is making efforts to support and educate retarded persons to their highest potential. Certainly the retarded have this right, but do they not also have the obligation not to have children they cannot support and who will become burdens to the taxpayer? Research is needed to identify and predict which young people, non-retarded as well as retarded, are incapable of assuming the responsibilities of parenthood.

Fletcher (1960) wrote that society should have the right to protect itself from the birth of children who would be a burden to the community.

As one means of helping to prevent juvenile delinquency, the prevention of parenthood for the retarded is important also. Glueck and Glueck (1950) found almost four times as many mentally retarded mothers in the families of delinquent boys as among nondelinquent controls, 32.8 percent compare to 9 percent. A psychiatrist at the Municipal Court in New York City wrote that many of the parents of delinquents were mentally subnormal and were barely able to look after themselves (Rittwagen, 1958).

CONCLUSION

In the light of the above facts and in agreement with the Brock Committee and the American Neurological Committee's investigations of eugenic sterilization, I would like to recommend that medical societies take a stand opposed to compulsory sterilization and take all possible measures to make family life sex education, genetic counseling, birth control, voluntary sterilization and abortion readily available to all. I would also recommend liberalizing state laws to allow marriage of retarded individuals.

Presser *(op. cit.)* concluded *Voluntary Sterilization: A World View,* as follows:

> Facilities must also be made available and the standards applied to individuals wanting sterilization must be flexible – or nonexistent – for sterilization to be truly voluntary and represent a real option for persons attempting to control their fertility.

To quote from Dr. Hall's (1965) study of abortion and sterilization:

> With or without the formality of legislation, some representative body of obstetricians [I would enlarge this to include psychiatrists, pediatricians, other physicians and geneticists and public health personnel] should establish an acceptable code of sterilization ethics. The obstetrician's obligation to provide abortion, sterilization and contraception is inadequately and inequitably met at the moment. The obstetricians of America must individually and collectively review these vital issues in an effort to establish a more uniformly humane birth control ethic *(op. cit.,* p. 531).

Physicians have the knowledge and prestige to accomplish this and by so doing may be able to reduce the rare of mental deficiency by one-third.

REFERENCES

Allen, M.: Sex related problems of the mentally retarded, Panel Discussion,

Region IV Meeting, Amer. Assoc. Mental Deficiency, Salt Lake City, Utah, Nov. 1, 1968.
Association for Voluntary Sterilization, Inc., Blue Cross-Blue Shield and Medicaid Insurance for Voluntary Sterilization, New York, 1972.
Bass, M. S. Attitudes of parents of retarded children toward voluntary sterilization, Eugenics Quarterly, March, 14(1):45-53, 1967.
, 1967. Attitudes of parents of retarded children toward voluntary sterilization, Eugenics Quarterly, March, 14(1):45-53.
Benda, C. E., Squires, N. D., Ogonik, M. J. and Wise, R.: Personality factors in mild mental retardation, Am. J. Ment. Defic., 68:104-5, 1963.
Birnbaum, M.: Eugenic sterilization, J. Am. Med. Assoc., March 18, 175:951-958, 1961.
Charles, D. C.: Adult adjustment of some deficient American children, Amer. J. Ment. Defic., 62:300-304, 1957.
Coakley, F. M.: Excerpts from Group Discussions in New Directions for Mentally Retarded Children, Josiah Macy Jr. Foundation, 16 West 46th St., New York.
Davis, J.: Voluntary sterilization, Urology. Reprint Assoc. for Voluntary Sterilization, 1970.
Dunn, L. M.: White House Conference on Mental Retardation, Proceedings, U. S. Govt. Printing Office, 104-105, 1963.
Fletcher, J. F.: Morals and Medicine, Boston, Beacon Press, 1960, p. 168.
Fujita, B., Wagner, N. N. and Pion, R. J.: Sexuality, contraception and the mentally retarded, Post. Grad. Med., May, 47(5):193-197, 1970.
Glueck, S. and E.: Unraveling Juvenile Delinquency, Commonwealth Fund.
Goodman, L., Budner, S. and Lesh, B.: The parents' role in sex education for the retarded, Mental Retardation, 9(1):43-45, 1971.
Hall, R. E.: Therapeutic abortion, sterilization and contraception, Am. J. Ob-Gyn., February, 91(4):516-532, 1965.
Hammar, S. L., Wright, L. S., and Jensen, D. E.: Sex education for the retarded adolescent, Clinical Pediatrics, November, 621-627, 1967.
Heber, R. and Garber, H.: An experiment in the prevention of cultural-familial mental retardation, (in press). Proceedings of the Second Congress of International Assoc. for Scientific Study of Mental Deficiency, Warsaw, Poland, Sept. 2, 1970.
Hobbs, N.: The human rights of the retarded, paper presented to the Kennedy Foundation Scientific Symposium, Oct. 16, 1971.
Jastak, J. F., MacPhee, H. M., Whiteman, M.: Mental Retardation: Its Nature and Incidence, Univ. of Del., 1963.
Johnson, B. S.: A study of cases discharged from Laconia State School from July 1, 1924, to July 1, 1934, Amer. J. Ment. Defic., 50:437-445, 1946.
Johnson, W. R.: Sex education and the mentally retarded, J. Sex Research, 5(3):179-182, 1965.
Joint Committee on Voluntary Sterilization, Departmental Committee on Sterilization Report, London, 1934.

Kagan, J.: American longitudinal research on psychological development, 1967.

Kemp, T.: Genetic-hygienic experiences in Denmark in recent years, Eugenics Review, November, 49:11-18, 1957.

Kennedy, R. J. R.: The social adjustment of morons in a Connecticut city, Governor's Commission to Study the Human Resources of the State of Conn., in collaboration with the Carnegie Institute of Washington, 1948.

Kratter, F. E.: A modern approach to mental deficiency, N. C. Med. J., 19(7):268-271, 1958.

Lancet, Editor, Jan., 1962.

Masland, R. L., Sarason, S. B. and Gladwin, T.: Mental Subnormality, Basic Books, New York, 1959.

Mattinson, J.: Marriage and Mental Handicap, paper presented at conference of National Institute of Child Health and Human Development, Hot Springs, Arkansas, Nov. 9, 1971.

Menninger, K.: Love Against Hate, Harcourt Brace, New York, 1942.

Mental Retardation: a report on Am. Med. Assoc. Conf. on Mental Retardation, Apr. 1964, J. Am. Med. Assoc., 191:3, Jan. 18, 1965.

Mickelson, P.: Can mentally deficient parents be helped to give their children better care?, Amer. J. Ment. Defic., 53:516-534, 1949.

Mitchell, S. B.: Results in family case work with feebleminded clients, Smith College Studies in Social Work, 18, 1947.

Muller, C. P., Health Insurance for Abortion Costs Perspectives, Planned Parenthood World Population, 515 Madison Ave., New York, October, 2:4, 1970.

Myerson, A.: Eugenical Sterilization, New York: Macmillan Co., 1936.

New Medica Materia, 1962. November, p. 19.

Paul, J.: The psychiatrist as public administrator; Case in point: State sterilization laws, Am. J. Orthopsychiatry, January, 38:(1):76-82, 1968.

Pavenstedt, E.: A comparison of the child-rearing environment of upper-lower and very low-lower class families, Am. J. Orthopsychiatry, 35(1):89-98, 1965.

Peck, J. R., & Stephens, W. B.: Marriage of young adult male retardates, Amer. J. Ment. Defic., 60(6):818-827, 1965.

Penrose, L. S.: Mental Defect. New York: Farrar Rinehart, 1934.

Planned Parenthood Institute, Human Sexuality and Mental Retardation, Philadelphia, 1971.

Pollock, M.: Mental disease among mental defectives, Amer. J. Ment. Defic., 49:477-480, 1945.

President's Task Force on the Mentally Handicapped, 1970.

Presser, Harriet B.: The role of sterilization in controlling Puerto Rican fertility, Population Studies, 23:343-361, 1969.

Presser, H. B.: Voluntary sterilization: A world view, Reports on Population/ Family Planning, The Population Council, Inc., 256 Park Ave., New York, 10017, July, No. 5, 1970.

Public Welfare Report, Dept. Public Welfare, Penna., Dec., p. 85, 1969.
Reed, E. W. and S. W.: Mental Retardation: A Family Study, Phila., W. B. Saunders, 1965.
Rittwagen, M.: Sins of Their Fathers, Boston: Houghton-Mifflin, 1958.
Sarason, S. B.: Psychological Problems in Mental Deficiency, New York, Harper, 1953.
Shafter, A. J.: Criteria for selecting institutionalized mental defectives for vocational placement, J. Ment. Defic., 61:599-614, 1957.
Shaw, C. H. and Wright, C. H.: The married mental defective: A follow-up study, Lancet, January 30, 273-274, 1960.
Sheridan, M. D.: The intelligence of 100 neglectful mothers, Brit. Med. J., 1:91, 1956.
Weintraub, P.: Sterilization in Sweden: Its law and practices, Amer. J. Ment. Defic., 56:364, 1951.
White House Conference on Mental Retardation, 1930.
Wolfensberger, W.: Will there always be an institution?, Mental Retardation, October, 9(5):14-20, 1971.
Wood, H. C.: Sex Without Babies, Philadelphia, Whitmore Publ. Co., 1967.
Zellweger, H.: Counseling in clinical medicine, reprint from Modern Medicine, October 9, 40-52, 1967.

Chapter IX

VIEWS OF A GENETICIST ON EUGENIC STERILIZATION

PAUL S. MOORHEAD

THERE are some dozen or so textbooks on human genetics commonly encountered and these have been written over the last thirty years or so. Nearly all have a small section in the back entitled *Eugenics* or the *Future of Genetics for Man.* Such chapters often consist of only a few pages since the author usually does not wish to deal with such a touchy subject. Eugenics does involve many types of controversy – legal, ethical and scientific and the dangers of racism are perhaps sufficient reason for avoiding any use of the term *eugenics* whatever. However, genetic facts do not stand or fall on the basis of our opinions or upon the basis of misrepresentation or perversion of principles.

We academicians cannot avoid being dragged through the back door into public discussions and decisions on these problems. Technological advances and invention have a way of opening doors and changing society's practice entirely apart from decision making by any deliberative body, whether of elected leaders or of scientists. Many people today believe that abortion ought to be available at will and a liberalization of laws on abortion has proceeded rapidly in recent years.

Our laboratory is involved indirectly in these decisions whether we choose to be or not. If after amniocentesis and examination of the chromosomes of cells from the amniotic fluid it is determined that the fetus carried by a pregnant woman is trisomic for a chromosome #21, the outcome of a full term pregnancy is a certainty. If not terminated, the result will be an idiot, a case of Down's syndrome. This is a genetic defect but in a way somewhat different from the problem of the classic single factor dominant or recessive trait. There are perhaps 2×10^6 or so *genes* in the human hereditary endowment and the dosage of this extra chromosome amounts to about ½ percent of the total genetic content. It matters very little whether an extra chromosome involving hundreds or a thousand genes or a single gene factor is the cause of dysgenesis. If the end result is diagnosable and defined, having known deleterious properties, the situation is then predictable. These known genetic conditions are numerous but would not constitute more than a portion of all the cases of embryonic malformation encountered.

Certain authors of genetic texts understand population genetics quite well, however the theoretical considerations that we may be able to apply do not necessarily embrace all the factors in man's world and unavoidably many assumptions are necessary. The validity of the usual graphs reproduced in texts showing the elimination of a dominant or recessive condition or elimination of the gene associated is not questioned but the applicability to humans is. From experience with caged Drosophila populations, in which one can precisely define the genotype of input, predictions are often not fulfilled.

The particular textbook, by Snyder,* which I used twenty years ago was most interesting to read again on this subject. Snyder's opinions have remained sound and valid and I would recommend his chapter to your attention. Speaking against the state laws on sterilization, Snyder noted that *The only conditions which are widely stated as grounds for sterilization are idiocy, imbecility, feeble mindedness, insanity, and epilepsy.* These are included in virtually all state statutes. He goes on to show that various sterilization laws in this country list some twenty-four classes of defect to which they are applicable and many of these are sterilizable in one or in two states. However, rarely if ever, does a statute designate as grounds for sterilization one of those specific defects which we definitely know to be a genetic entity! On the contrary, broad classes of defect about which we can know very little are usually the focus of attention in most of these laws. Dr. Tarjan† has effectively developed this point, and I wish to reemphasize it.

*Snyder, L. H.: Medical Genetics, Duke University Press, 1941.

†See Chapter II.

When and if one is dealing with a clearly identifiable genetic trait, then sterilizations could conceivably reduce the numbers of individuals affected. What is not appreciated is that conditions such as *feeblemindedness* lump together cases having no genetic component with those in which there may be a partial or complete genetic contribution. Even in the latter case multiple factors are suspected of being involved so that the effect of selection cannot be determined. Apart from complexities of dealing with the elimination of genes from populations, error is compounded by the use of unfounded criteria of measurement of mental capacity. It is clear to most psychologists and others working with the problem that concepts of native intelligence cannot be well measured and the concept of IQ is only a poor approximation.

If we use laboratory rats and breed only from the upper part of the bell-shaped curve of distribution based on mental performance, we will obtain another bell-shaped curve in the offspring. The center of this new curve of performance will be displaced to a point corresponding roughly with the point of origin of the parents from the original population distribution. But these *classes* are not genetic entities. We are not speaking of just one thing when we use terms such as *feebleminded* or *subnormal;* these are statistical conglomerates, not biologically true types. Thus in any slice of the population there is available an enormous variability, a capacity for reproducing another spread of distribution of ability in the next generation.

It is also very important to support the point that those more severe conditions which might possibly justify some sort of social intervention are exactly the ones for which we would be unable to obtain an informed consent.

Additionally, these cases are precisely the ones least likely to be capable or effective in having viable offspring, either because of being sequestered by society or because of associated biological incapacity.

Altogether, these arguments diminish the eugenic return which society might theoretically obtain by any sterilization of affected persons. This returns us to the individual or to the individual family and a consideration of the burdens which a severe condition places upon that family. The immediate personal and ethical questions involved in managing subnormal persons remain.

It is encouraging, at least in my personal opinion, that the means for termination of unwanted pregnancies has become safe, reliable and available. Rather than sterilization, such approaches to prevent individual tragedy seem to be a much more promising direction for our future.

We must not reject forever the possibility of the application of eugenic principles at some time in man's future. Our personal cowardice may very well lead most of us to rationalize inaction respecting advocacy of any kind of social recommendations. What we choose to adopt as a free people in a supposedly democratic society must await appropriate scientific understanding.

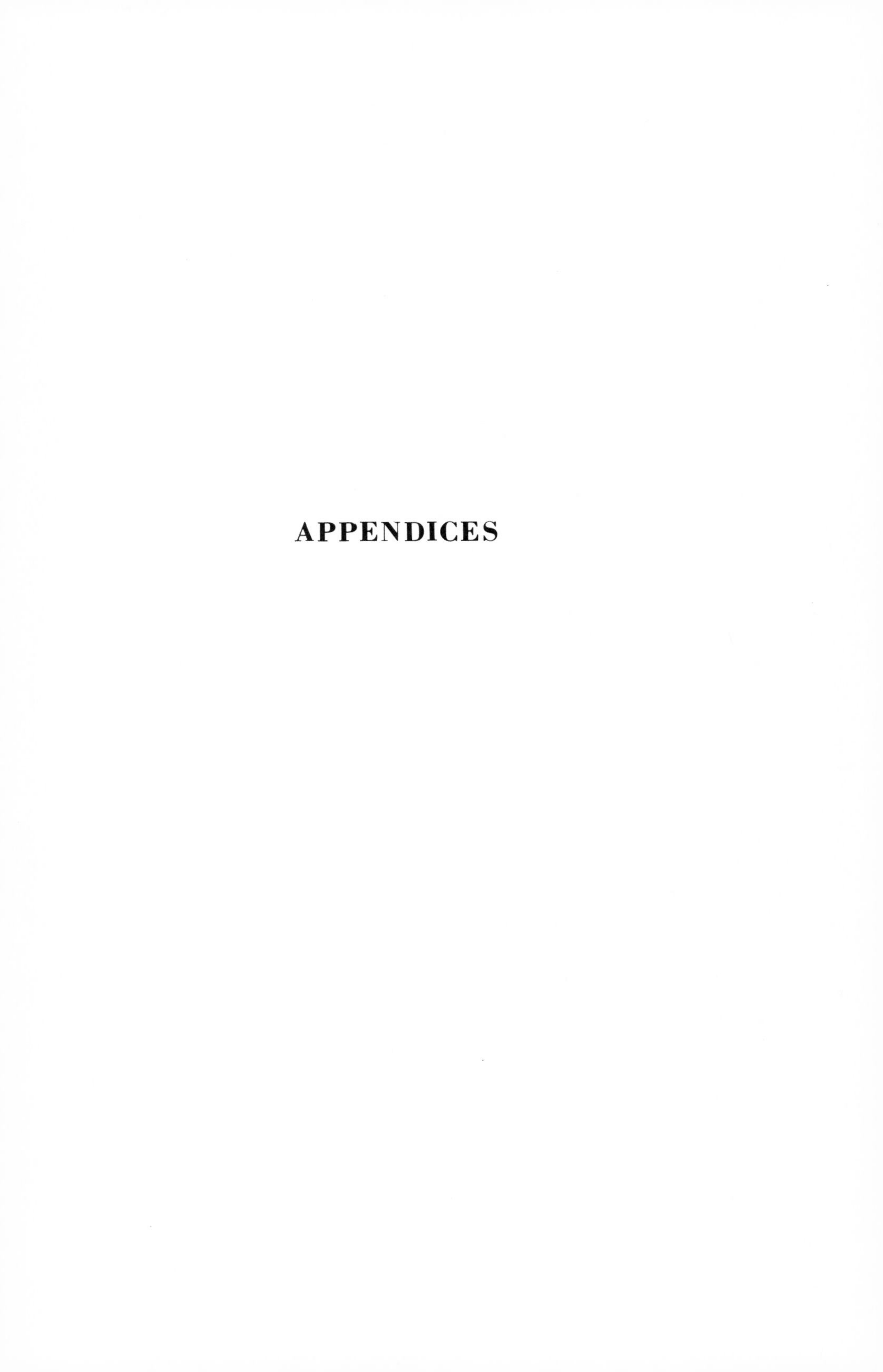

APPENDICES

Appendix 1

ANNUAL STERILIZATIONS PERFORMED UNDER ST
TOGETHER WITH CUMUL

	ALABAMA*	ARIZONA	CALIFORNIA	CONNECTICUT	DELAWARE	GEORGIA*	IDAHO	INDIANA	IOWA	KANSAS*	MAINE	MICHIGAN	MINNESOTA	MISSISSIPPI
TOTALS UP TO 1/1/43	224	20	16,553	457	641	190	14	1,231	493	2,706	217	2,388	2,111	542
OPERATIONS IN														
1943		–	459	31	17	97	–	55	58	88[1]	1	78	46	4
1944		–	387	2	27	73	–	74	46	57	–	107	34	8
1945		–	436	1	20	90	–	122	52	32	2	86	13	9
1946		–	480	–	29	68	–	192	45	71	9	75	2	19
1947		–	401	6	15	24	–	89	70	29	3	117	1	11
1948		–	326	8	34	94	–	77	127	18	3	131	4	3
1949		1	381	10	19	167	–	49	165	–	3	88	8	–
1950		–	275	2	13	226	–	71	113	24	6	72	12	–
1951		–	150	3	7	200	–	60	178	–	4	65	15	–
1952		–	39	5	33	279	5	37	70	–	5	81	16	–
1953		–	23	5	4	246	8	55	85	–	8	103	13	–
1954		–	27	2	–	207	4	85	72	–	17	71	10	–
1955		1	25	3	–	261	2	94	47	–	26	61	9	–
1956		8	23	7	12	268	–	34	69	–	1	27	19	6
1957		–	13	2	8	268	–	29	48	–	5	47	12	13
1958		–	13	2	1	142	–	17	21	–	4	39	8	22
1959		–	12	1	0	112	–	7	14	–	4	27	6	15
1960		–	18	2	9	148	1	14	28	–	2	18	5	13
1961		–	24	–	9	60	2	9	51	7[3]	3	46	5	3
1962		–	26	5	22	57	–	11	28	–	2	26	1	8
1963		–	17	3	25	7	2[4]	12	30	–	1	33	–	7
TOTALS UP TO 1/1/64	224	30	20,108	557	945	3,284	38	2,424	1,910	3,032	326	3,786	2,350	683

ATE EUGENIC STERILIZATION LAWS (1943-63)
ATIVE GRAND TOTALS

MONTANA*	NEBRASKA*	NEW HAMPSHIRE	NEW YORK*	NORTH CAROLINA	NORTH DAKOTA*	OKLAHOMA	OREGON	PUERTO RICO*	SOUTH CAROLINA	SOUTH DAKOTA	UTAH	VERMONT	VIRGINIA	WASHINGTON*	WEST VIRGINIA	WISCONSIN	GRAND TOTAL
208	530	468	42	1,346	628	553	1,597		57	643	310	225	4,472	685	47	1,372	40,970
5	79	24		152	43	–	57		17	3	69	3	203		1	48	1,638
–	23	19		107	13	–	24		3	7	16	4	105		–	47	1,183
2	16	20		117	21	–	21		–	10	22	8	178		–	58	1,336
15	13	9		105	29	–	49		–	24	56	10	152		–	24	1,476
6	7	6		139	17	–	30		–	46	58	–	122		–	35	1,232
5	20	11		186	33	–	43	986[2]	4	12	16	1	134		–	46	1,336
8	16	14		249	23	–	32		7	3	14	–	215		–	28	1,500
–	19	17		295	23	–	60		12	5	34	1	204		–	42	1,526
3	19	23		375	42	–	42		10	4	30	–	207		–	22	1,459
2	37	18		326	22	2	72		3	4	46	–	153		–	12	1,267
–	27	21		270	37	–	59		7	8	16	–	169		–	16	1,180
2	23	8		300	–	–	25		8	6	13	–	171		14	14	1,079
–	15	4		289	16	1	28		30	2	9	–	111		33	–	1,067
–	8	8		216	14	–	38		43	2	15	–	87		3	1	909
–	5	8		305	14	–	23		34	4	12	1	128		–	7	986
–	4	–		318	9	–	14		11	3	5	–	114		–	–	747
–	10	1		260	22	–	40		6	1	7	–	69		–	–	614
–	6	–		234	12	–	19		7	1	16	–	61		–	8	622
–	10	–		248	16	–	24		3	–	–	–	39		–	2	561
–	5	–		220	15	–	21		6	–	–	–	29		–	6	488
–	10	–		240	–	–	23		9	1	–	–	39		–	8	467
256	902	679	42	6,297	1,049	556	2,341		277	789	772	253	7,162	685	98	1,823	63,678

Appendix 1

Statistical Footnotes

All of these figures are taken from the published compilations of Birthright, Inc. (1943-49) and its successor, the Human Betterment Association of America, Inc., (HBAA), now called the Association for Voluntary Sterilization, Inc. (1950-63). In some cases, these figures do not correspond to the official figures given at the end of some of the state sections, but no effort to mix the two sets of figures has been attempted. The 1963 compilation was the last AVS collection of reported state sterilizations, after 1963 the figures decline drastically and are statistically insignificant.

Author's Notes

* The Alabama law was enacted in 1919, but has been unenforced since the Alabama Supreme Court handed down an adverse advisory opinion in 1935 holding unconstitutional a proposed new sterilization bill. The figure of 224 is for the period up to the time of the decision. Alabama authorities claim that no operations under the law have been performed since that time.

* * *

* Kansas repealed its law in 1965.

* * *

* The New York sterilization law of 1912 was declared unconstitutional by the Supreme Court, Albany County, in 1918 and affirmed by the Appellate Division of the Supreme Court of the State of New York for the Third Department in 1918. Pending decision by the state's highest court, the Court of Appeals, the state legislature repealed the law in 1920. New York's total of forty-two is for the period up to the time of this litigation.

* * *

* North Dakota repealed its law in 1965.

* * *

* The Commonwealth of Puerto Rico repealed its 1937 law in 1960.

* * *

* Montana repealed its law in 1969; Nebraska repealed its law in 1969; Georgia repealed its law in 1970.

* * *

* Washington's general eugenic sterilization statute, enacted in 1921, was declared unconstitutional by the state Supreme Court in 1942. Up to that time, 685 operations were performed under the act. An earlier 1909 punitive sterilization statute was held constitutional by the state Supreme Court in 1912 and is still currently part of the present Revised Code of Washington (1950), but no operations have ever been reported under this act.

* * *

1. One of the six state institutions in Kansas failed to report.
2. This figure seems to be a misprint, since the Commonwealth of Puerto Rico Department of Health reports that the law was inactive from 1946 to its repeal in 1960, and that a total of ninety-six operations were performed under the act, not 986, from 1937 until 1946, according to its records. HBAA agreed that the figure of 986 might have been inaccurate insofar as the law is concerned, but it *may* have included voluntary sterilizations.
3. The Kansas State Department of Social Welfare disputes this figure, claiming that no operations were performed under the law from 1950 until the law's repeal in 1965.
4. The Idaho Administrator of Health disputes the figures for the

period 1961-63, claiming one operation in 1963, and none for the other two years.

Appendix 2

PERIODIC CUMULATIVE GRAND TOTALS OF PERSONS STERILIZED UNDER STATE LAWS ACCORDING TO THE CATEGORIES COVERED BY THE LAWS, AND BY SEX, GATHERED FROM VARIOUS SOURCES

FOR THE PERIOD 1907 UP TO --	Mentally Ill	Mentally Retarded	MR % of C.G.T.	Epileptic	"Others"	Male	Female	California Total Alone	Cumulative Grand Total	Calif. % of C.G.T.
Jan. 1, 1921	2,700	403	12.5		130[1]	1,853	1,380	2,558	3,233[2]	80
July 1, 1925						3,307	2,937	4,636	6,244[3]	74
Jan. 1, 1929	6,246	2,938	31.7	55	21[4]	4,624	4,845	6,298	9,522[5]	66
Jan. 1, 1932						5,613	6,532	7,548	12,145[6]	62
Jan. 1, 1935						8,573	11,363	9,931	20,021[7]	49.6
Jan. 1, 1938						11,628	16,241	12,180	27,869[8]	44
Jan. 1, 1941	18,552	16,622	46		704	14,900	20,978	14,568	35,878[9]	41
Jan. 1, 1946	21,311	22,153	49		1,663	18,830	26,297	17,835	45,127[10]	39
Jan. 1, 1951	23,466	26,858	51		1,909	21,250	30,983	19,698	52,233[11]	38
Jan. 1, 1956	26,047	30,101	51.6		2,137	23,368	34,917	19,962	58,285[12]	34
Jan. 1, 1961	27,592	32,287	52		2,283	24,474	37,688	20,041	62,162[13]	32
Jan. 1, 1964	27,917	33,374	52.4		2,387	24,716	38,962	20,108	63,678[14]	31.6

ANNUAL AVERAGES OF OPERATIONS FOR THE PERIODS

1907 to 1921	(230)
1921 to 1930	(849)
1930 to 1941	(2,273)
1941 to 1951	(1,636)
1951 to 1961	(993)
1961 to 1964	(505)

Appendix 2

Footnotes

1. Came under the category of *criminalistic,* as used by Laughlin.
2. Laughlin, *ES in the U.S.,* p. 96.
3. Laughlin, *ES: 1926,* p. 60.
4. Included sixteen criminals and five undesignated, according to Brown.
5. Frederick W. Brown, Eugenic Sterilization in the United States: its present status, 149 *The Annals* 22 (Part III. May 1930) p. 30-32. Brown noted discrepancies in his cumulative totals for the sexes as against the totals for the various categories due to incomplete reporting on the part of some states, and the 9,522 cumulative total also includes fifty-three operations not differentiated according to sex.
6. Landman, *Human Sterilization,* p. 289.
7. ANA Report, *ES,* p. 8-20. Of the cumulative grand total, Montana's eighty-five reported operations were not designated according to sex.
8. Annual compilation for 1938, Human Betterment Foundation, Pasadena, California.
9. Annual compilation for 1941, Human Betterment Foundation. The category *others* as employed by the HBF included the epileptic and other classes covered by the statutes.
10. Annual compilation for 1946, Birthright, Inc., Princeton, New Jersey.
11. Annual compilation for 1951, Human Betterment Association of America, Inc. of New York City (Publication No. 5), *Sterilizations Reported to January 1, 1951.*
12. Annual compilation for 1956, HBAA, *Sterilizations Reported in the United States to January 1, 1956.*
13. Annual compilation for 1961, HBAA, *Sterilizations Reported in the United States Through December 31, 1960.*
14. Thirty-fifth annual compilation (final one issued), Human Betterment Association for Voluntary Sterilization, Inc., *Sterilizations Performed Through December 31, 1963 Under*

U.S. State Sterilization Statutes Primarily for the Institutionalized Mentally Ill and Mentally Deficient.

APPENDIX 2A

CUMULATIVE STATE STERILIZATIONS, 1907-1970

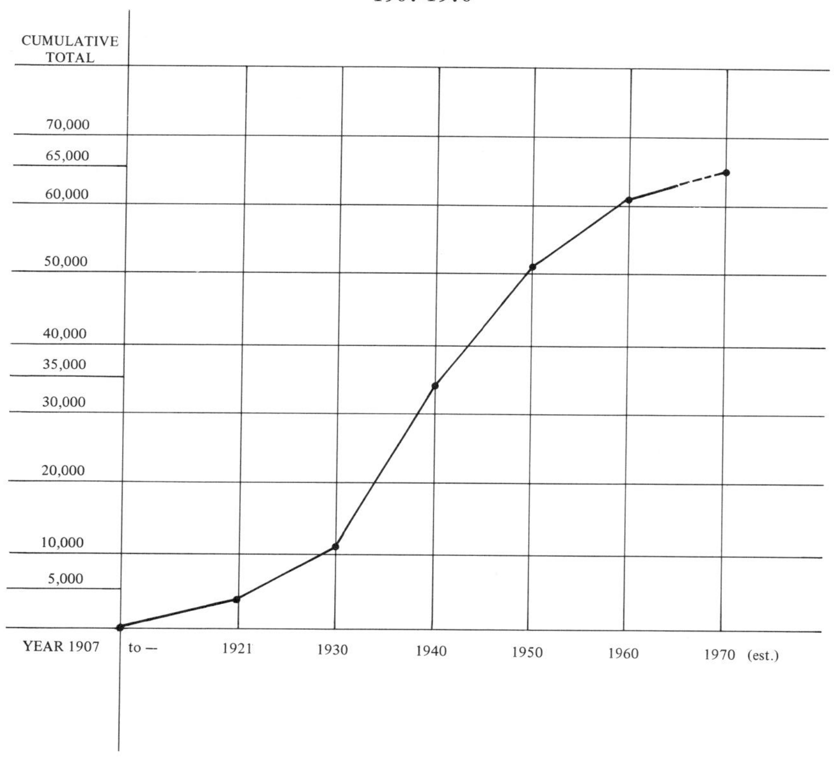

Appendix 3

STATE AND FEDERAL COURT DECISIONS REGARDING STATE STERILIZATION LAWS AND VARIOUS OTHER STATE DECISIONS REGARDING VOLUNTARY AND COMPULSORY STERILIZATION

1. *State v. Feilen,* 70 Wash. 65, 126 Pac. 75 (1912).
2.* *Smith v. Board of Examiners,* 85 N.J.L. 46, 88 Atl. 963 (1913).
3.* *Davis v. Berry,* 216 Fed. 413 (S.D. Iowa 1914), *rev'd for mootness, Berry v. Davis,* 242 U.S. 468 (1916).
4.* *Haynes v. Lapeer Circuit Judge,* 201 Mich. 138, 166 N.W. 938 (1918).
5.* *In re Thompson,* 103 Misc. 23, 169 N.Y. Supp. 638 (Sup. Ct. 1918), *aff'd mem. sub. nom., Osborn v. Thompson,* 185 App. Div. 902, 171 N.Y. Supp. 1094 (3d Dep't 1918).
6.* *Mickle v. Henrichs,* 262 Fed. 687 (D. Nev. 1918).
7.* *Cline v. Oregon State Board of Eugenics* (law declared unconstitutional by the Circuit Court for the County of Marion, Dec. 13, 1921; the state did not appeal to the state Supreme Court).
8.* *Williams v. Smith,* 190 Ind. 526, 131 N.E. 2 (1921).
9. *Buck v. Bell,* 143 Va. 310, 130 S.E. 516 (1925), *aff'd,* 274 U.S. 200 (1927).
10. *Smith v. Command,* 231 Mich. 409, 204 N.W. 140 (1925).
11. *In re Salloum,* 236 Mich. 478, 210 N.W. 498 (1926).
12.† *State ex rel. Smith v. Schaffer,* 126 Kan. 607, 270 Pac. 604 (1928).
13.**† *Davis v. Walton,* 74 Utah 80, 276 Pac. 921 (1929).
14.† *In re Clayton,* 120 Nebr. 680 234 N.W. 630 (1931).
15.† *State v. Troutman,* 50 Idaho 673, 299 Pac. 668 (1931).
16.* *Brewer v. Valk,* 204 N.C. 186, 167 S.E. 638 (1933).
17.† *In re Main,* 162 Okla. 65, 19 P.2d 153 (1933).
18.* *In re Opinion of the Justices,* 230 Ala. 543, 162 So. 123 (1935).

19. *Garcia v. State Dep't of Institutions,* 36 Cal. App. 2d 152, 97 P.2d 264 (1939).
20.* *In re Hendrickson,* 12 Wash. 2d 600, 123 P.2d 322 (1942).
21.* *Skinner v. Oklahoma ex rel. Williamson,* 189 Okla. 235, 115 P.2d 123 (1941), *rev'd on other grounds,* 316 U.S. 535 (1942), *remanded on issue of severability,* 195 Okla. 106, 155 P.2d 715 (1945), entire act held unconstitutional by Okla. Supreme Court.
22.† *Cavitt v. Nebraska,* 182 Neb. 712, 157 N.W. 2d 171 (1968); 183 Neb. 243, 159 N.W. 2d 566; 393 U.S. 1078, 396 U.S. 996 (1970), dismissed as moot due to eventual repeal of statute by Nebraska Legislature in 1969.

Sterilization Decisions in States Without Sterilization Laws

1. *Ex parte Eaton,* Baltimore City Circuit Court (Nov. 10, 1954), reported in *Baltimore Daily Record* (Nov. 12, 1954).
2. *In re Simpson,* 180 N.E. 2d 206 (Ohio P.Ct. 1962).
3. *Frazier v. Levi,* 440 S.W. 2d 393 (Ct.Civ.App.Tex. 1969).
4. *Holmes v. Powers,* 439 S.W. 2d 579 (Ky. 1969).

State Court Decisions Regarding Voluntary Sterilization and Public Policy

1. *Christensen v. Thornby,* 192 Minn. 123, 255 N.W. 620 (1934).
2. *Wiley v. Wiley,* 59 Cal. App. 2d 840, 139 P.2d 950 (1943). Incidental issue in this case.
3. *Shaheen v. Knight,* 6 Lyc. 19, 11 Pa. D. & C.2d 41 (1957). Case noted, 19 *U. Pitt. L. Rev.* 802 (1958).
4. *People v. Blankenship,* 16 Cal. App. 2d 68, 61 P.2d 352 (1936). Involved the validity and reasonableness of vasectomy as a condition of probation in rape case.
5. *Corman v. Anderson,* memorandum trial court decision No. 701.588, Superior Court, Los Angeles, California (Nov. 16, 1960).
6. *Ball v. Mudge,* 64 Wash. 2d 247, 391 P.2d 201 (1964). Case noted, 6 *Ariz. L. Rev.* 318 (1965).
7. *Jessin v. County of Shasta,* 79 Calif. Rptr. 359 (1969).

Cases Regarding Consent to a Voluntary Sterilization

1. *Kritzer et al. v. Citron et al.,* 101 Cal. App. 2d 33, 224 P.2d 808 (1950).
2. *Danielson v. Roche et al.,* 109 Cal. App. 2d 832, 241 P.2d 1028 (1952).
3. *Maercklein et al. v. Smith,* 129 Colo. 72, 266 P.2d 1095 (1954). Discussed in 159 *J.A.M.A.* 1149 (1955).

Cases Regarding Voluntary Sterilization Operations

1. *Doerr v. Villate,* 220 N.E. 2d 767 (Ill. 1966). Unsuccessful vasectomy.
2. *Bishop v. Byrne,* 265 F. Supp. 460 (D.C., W.Va., Mar. 21, 1967). Unsuccessful sterilization operation.
3. *Custodio v. Bauer,* 59 Cal. Rptr. 463 (Cal., May 24, 1967; rehearing denied June 13, 1967.) Pregnancy after sterilization operation is cause for suit.
4. *Lane v. Cohen,* 201 So. 2d 804 (Fla., Aug. 8, 1967). Pregnancy after vasectomy.
5. *Vilord v. Jenkins,* 226 So. 2d 345 (Fla. App., 1969). Pregnancy after female sterilization.
6. *Jackson v. Anderson,* 230 So. 2d 503 (Fla. App., 1970). Pregnancy after female sterilization.

* * *

FOOTNOTES

* The law was invalidated on any of several Constitutional grounds (*i.e.,* as violative of due process of law, equal protection of the laws (*class legislation*), cruel and unusual punishment, or a bill of attainder).

** The statue was upheld, but the Court held that there was insufficient evidence to support a finding that the inmate was the probable potential parent of similarly afflicted offspring (he had been found practicing sodomy with other inmates), and therefore reversed and remanded the sterilization order.

† The statute was upheld against constitutional attack, relying in part on the decision by the U.S. Supreme Court in *Buck v. Bell* (1927). All of the remaining *unmarked* cases in the list are also cases where the sterilization statute was upheld against such attack.

Appendix 4

CURRENT LIST OF STATE STERILIZATION LAWS BY STATUTORY AND CODE CITATIONS

ALA.	ALA. CODE tit. 45, S 243 (Recompiled 1958).
ARIZ.	ARIZ. REV. STAT. ANN. SS 36-531 through 36-540 (1956).
CAL.	CAL. CODE ANN. PENAL CODE S 645 (West's 1970), S 2670 (West's 1970); WELFARE & INSTITUTIONS CODE S 6624 (1970 P.P.), repealed by Stats. 1967 (1970 P.P.) and replaced by S 7254 (1970 P.P.)
CONN.	CONN. GEN. STAT. ANN. S 17-19 (1970 P.P.), amended, 1965, 1967, 1969.
DEL.	DEL. CODE ANN. tit. 16, SS 5701 through 5705 (1953), ch. 57 (see 1970 P.P. for changes).
GA.	GA. CODE ANN. Compulsory eugenic sterilization law repealed in 1970. Voluntary Sterilization Act passed in 1966 and revised in 1970. SS 84-931 through 84-936 (1971 P.P.).
IDAHO	IDAHO CODE ANN. SS 66-801 through 66-812 (1949), SS 66-801, 66-806 (Supp. 1969), SS 66-803 (Supp. 1971). Also, see SS 39-101(15) (1971 Cum.Pock.Supp.), and SS 39-3901 through 39-3909 (Sterilization) in (1971 C.P.Supp.).
IND.	IND. STAT. ANN. SS 22-1601 through 22-1618 (Burns, Repl. vol. 1964 and 1971 CPS). Also, see 22-5003, 22-5006, 22-5007, 22-5008,

	22-5009, (1971 C.P.Supp.), 22-5032 (Burns 1964).
IOWA	IOWA CODE ANN. SS 145.1 through 145.22 (1949), SS 145.1 through 145.5, 145.9, 145.11 through 145.19 (1970 P.P.)
ME.	ME. REV. STAT. ANN. tit. 34, SS 2461 through 2468 (1965).
MICH.	MICH. COMP. LAWS ANN. SS 720.301 through 720.310 (1968).
MINN.	MINN. STAT. ANN. SS 256.07 through 256.10 (1971 P.P.).
MISS.	MISS. CODE ANN. SS 6957 through 6964 (Recompiled 1952), see SS 6957 (1971 Supp.).
MONT.	MONT. REV. CODE ANN. SS 69-6401 through 69-6406 (Second Replacement Vol. 4, 1970). Voluntary law replaced compulsory act repealed in 1969 Laws). SS 69-6402 (1971 C.P.Supp.).
N.H.	N.H. REV. STAT. ANN. SS 174:1 through 174:14 (Replacement ed. 1964).
N.C.	N.C. GEN. STAT. ANN. SS 35-36 through 35-57 (1966 Replacement Vol. and 1969 Cum.Supp.). Volunatry sterilization law, SS 90-271 through 90-275 (1965), SS 90-271 and 272 (1971 Cum.Supp.).
OKLA.	OKLA. STAT. ANN. titl. 43A, SS 341 through 346 (1954 and 1971 P.P.). Also, see SS 11

through 14 (1954 and Supp. 1965), 32-33 (1954 and 1971 P.P.).

ORE.	ORE. REV. STAT. SS 436-010 through 436-150 (1969), Annotations (1953, and Cum.Supp. 1962). Also, see SS 435.305 (1969), Voluntary Sterilization.
S.C.	S.C. CODE ANN. SS 32-671 through 32-680 (1962), SS 32-679 (1970 Cum.Supp.).
S.D.	S.D. COMP. LAWS ANN. SS 27-11-1 through 27-11-6 (1967), sterilization of mentally ill persons; SS 27-17-1 through 27-17-34 (1967), sterilization mentally retarded persons; also see, SS 25-1-16 (1967), issuance of marriage licenses; SS 27-13-7 (1967), consent requirements; SS 27-15-15 (1967), list of mentally retarded filed with clerk of courts.
UTAH	UTAH CODE ANN. SS 64-10-1 through 64-10-14 (1961), SS 64-10-6 and 64-10-7 (Supp. 1965).
VT.	VT. STAT. ANN. SS 18-8701 through 18-8704 (1968 Replacement Ed.).
VA.	VA. CODE ANN. SS 37.1-156 through 37.1-171 (Replacement Vol. 1970); SS 32-423 through 32-427 (Replacement Vol. 1969), 32-424 (1971 Cum.Supp.), voluntary sterilization law.
WASH.	WASH. REV. CODE ANN. S 9.92.100 (1961). This is a 1909 punitive sterilization law, aimed at certain sex offenders and habitual criminals, that is still on the books but was presumably never used.

W. VA.	W. VA. CODE ANN. SS 16-10-1 through 16-10-7 (1966), SS 16-10-1 (1971 Cum.Supp.).
WIS.	WIS. STAT. ANN. S 46.12 (1957, and 1971 P.P.).

STATES WITH NO STERILIZATION LAWS BUT WITH PROHIBITIONS ON SUCH OPERATIONS

N.J.	N.J. STAT. ANN. S 30:11-9 (1964).
TEX.	TEX. REV. CIV. STAT. ANN. art. 3174b-2 (Vernon's Supp. 1968 and 1971 P.P.).

STATES WITH NO STERILIZATION LAWS BUT WHERE EQUITY OR PROBATE POWERS WERE USED TO ORDER STERILIZATIONS

MD.	MD. CODE ANN. art. 16, S 132 (1966 Replacement Vol.), Non Compos Mentis provision, repealed by Acts 1969 (1971 C.S.).
OHIO	OHIO REV. CODE ANN. tit. 51, S 5125.30 (Baldwin, 1970) relating to: Procedure for commitment of feebleminded when institutions crowded; Procedure when unable to receive patients.

STERILIZATION LAWS IN CANADIAN PROVINCES

ALBERTA	ALBERTA REV. STAT. c. 311 (1955).
BRIT. COLUM.	BRIT. COLUM. REV. STAT. c. 353 (1960), S 2 amended by c. 29, S 47 (Statutes 1964).

EUGENIC MARRIAGE LAWS WITH STERILIZATION PROVISIONS

NEB.	NEB. REV. STAT. S 42-102 (Reissue 1960, and Cumulative Supp. 1963).

N.C.	N.C. GEN. STAT. ANN. S 51.12 (Recompiled 1950).
S.D.	S.D. CODE S 30.0413 (1939).
UTAH	UTAH CODE ANN. S 30-1-2-(1) (1953), S 30-1-2.1 (Supp. 1965).

STATES WITH NO STERILIZATION LAWS BUT WITH STATUTES DEALING WITH STERILIZATION AND HEALTH INSURANCE COVERAGE

TENN.	TENN. CODE ANN. S 56-2933 (1971). Contract cannot exclude sterilization procedures.

INDEX

C

D

Q

R

S

T

Y

Z